HSÜN YÜEH AND THE MIND OF LATE HAN CHINA

PRINCETON LIBRARY OF ASIAN TRANSLATIONS

*Advisory Committee: Cyril Birch, Eugene Eoyang,
F. W. Mote, A. W. Plaks*

CH'I-YÜN CH'EN

Hsün Yüeh and the Mind of Late Han China

A TRANSLATION OF THE
SHEN-CHIEN
WITH INTRODUCTION
AND ANNOTATIONS

Princeton University Press
PRINCETON, N.J.

Copyright © 1980 by Princeton University Press
Published by Princeton University Press, Princeton, New Jersey
In the United Kingdom: Princeton University Press,
Guildford, Surrey

All Rights Reserved
Library of Congress Cataloging in Publication Data will be
found on the last printed page of this book

Publication of this book has been aided by
The Andrew W. Mellon Foundation
This book has been composed in Monophoto Baskerville
by Asco Trade Typesetting Limited, Hong Kong

Clothbound editions of Princeton University Press books
are printed on acid-free paper, and binding materials are
chosen for strength and durability.

Printed in the United States of America
by Princeton University Press, Princeton, New Jersey

TO
Yvonne Yuen-Han Chen

PREFACE

I wish to reiterate my indebtedness to all those whose influence on my intellectual encounter with Hsün Yüeh has been gratefully acknowledged in my first book, *Hsün Yüeh (A.D. 148–209): The Life and Reflections of an Early Medieval Confucian* (Cambridge University Press, 1975).

In publishing this second book on Hsün Yüeh, my special thanks go to Professor Ying-shih Yü (formerly of Harvard and now at Yale), Professor Erik Zürcher (Sinologisch Institut, Leiden), Professor Richard Mather (University of Minnesota), and Professor Frederick Mote (Princeton University). But for their encouragement and support, the completion of my work would have been delayed for a much longer time. My thanks also go to the editors, particularly R. Miriam Brokaw, at Princeton University Press for their expert handling of the manuscript, to the Monumenta Serica Institute for its permission for the inclusion of a revised version of my article in the *Monumenta Serica*, Vol. xxvii (1968), pp. 208–232, as part of the Introduction in the present book, to Professor David Knechtges for reading and correcting my translation, and to Gloria Leitner and Paul Pao-k'ang Ma for their assistance.

The reviews of my first book on Hsün Yüeh remind me how little we know about this historian-philosopher or the period in which he lived. It was precisely to remedy this situation that I tried to reconstruct the political events and the social, institutional, and intellectual changes of the Later Han period in greater detail in that book than would normally have been called for in a biographical study. Although the effort may have saved Hsün Yüeh "from the obscurity to which he has been consigned by the dismissive clichés of generations of historians" (Ian McMorran's review, *Times Literary Supplement*, May 7, 1976), it does not succeed in rescuing Hsün Yüeh's thought or the intellectual history of the Later Han from the scholarly neglect from which it has suffered.

This is partly my fault. In studying the life and times of

PREFACE

Hsün Yüeh, I found my attention drawn to the historical developments leading to the breakdown of the Later Han imperium and the subsequent effort for its restoration, to the neglect of the long-range intellectual significancy of these developments. Even in my discussion of Hsün Yüeh's writings and thought, the emphasis was mainly on their historical value as the product of a period of upheaval and change, rather than on their intrinsic merits. Somehow, I took the meaning of *chien* (reflection) too literally and portrayed Hsün Yüeh merely as a "mirror" reflecting a sequence of juxtaposed events, with little inner structure of its own. As a result, the Hsün Yüeh that emerges in that book remains a minor historian-thinker who happened to witness an epoch-making historical drama in his lifetime. In this respect, I have done Hsün Yüeh grave injustice.

It is to right a wrong that I now present to the English-reading public my translation of Hsün Yüeh's *Shen-chien*, with a long introductory essay on the philosophical import of his ideas. This, I hope, will help the reader to judge for himself the accomplishment of Hsün Yüeh as a thinker and to better appreciate his words of reflection, in addition to providing the insight that one may thus gain into the mind of late Han China. Some of the ideas in this book have been presented for discussion at the 1977-1978 Berkeley Seminar on Confucianism. I am grateful to members of that Seminar, especially Professor Tu Wei-ming (organizer) and Ms. Kathy Darcy (my doctoral student), for their comments.

CONTENTS

PREFACE	vii
Introduction	
I. Hsün Yüeh and the Mind of Late Han China	3
II. Textual Problems of Hsün Yüeh's Works: The *Han-chi* and the *Shen-chien*	54
III. Selections from Hsün Yüeh's *lun* (Discourses) in the *Han-chi*	80
Translation of the *Shen-chien* (Extended Reflections)	
1. Essence of Government (*Cheng-t'i*)	103
2. Current Affairs (*Shih-shih*)	126
3. Common Superstitions (*Su-hsien*)	150
4. Miscellaneous Dialogues (*Tsa-yen*), I	163
5. Miscellaneous Dialogues (*Tsa-yen*), II	179
BIBLIOGRAPHY	199
INDEX-GLOSSARY	209

INTRODUCTION

I. Hsün Yüeh and the Mind of Late Han China

Between A.D. 196 and 205, Hsün Yüeh 荀悅 (A.D. 148–209), Custodian of Secret Archives (*Mi-shu chien*) at the court of the last Han ruler, produced two important works, the *Han-chi* and the *Shen-chien*. The *Han-chi* is a chronicle of the Former Han dynasty (206 B.C. to A.D. 9), employing both implicitly and explicitly the "praise and blame" (*pao-pien*) historiographical principle to present some solemn lessons (*chien* 鑑) on the rise and fall of the regime. The *Shen-chien* is a philosophical work (*tzu*) in the traditional Chinese sense, in which Hsün Yüeh extended (*Shen* 申) his reflections (*chien*) on the issues of his time, ranging from politics to vulgar superstitions.[1]

These works represent the observations and ideas of an eminent historian-philosopher living in a time of great political upheaval and dramatic intellectual reorientation. As a young scholar, Hsün Yüeh had been involved in the *ch'ing-i* (pure criticism) movement of protest by the Confucian idealists against political corruption and moral degeneration at the Later Han court. In his mature years, he saw the court and empire crumble in ever-worsening civil strife. Subsequently, as a prominent official at the court of the figurehead Later Han Emperor Hsien (r. A.D. 190–220), he participated in a fruitless effort to restore the imperium. But it was too late. The Later Han dynasty came to an end in A.D. 220. What followed was a

[1] For a detailed study of Hsün Yüeh's life and times, with special emphasis on the major political events and important social and institutional changes preceding and following the fall of the Later Han dynasty, see Chi-yun Chen, *Hsün Yüeh (A.D.148–209): The Life and Reflections of an Early Medieval Confucian* (Cambridge University Press, 1975), hereafter abbr. as *Hsün Yüeh*. The present chapter will examine the philosophical meaning of Hsün Yüeh's works and his contribution to China's evolving intellectual legacy.

INTRODUCTION

long period of political disunity and barbarian conquest. Confucianism, which for centuries had been the orthodox teaching of the Han imperium, lost its privileged status; its usefulness as an ideology and the validity of its moral doctrines were increasingly questioned by disillusioned statesmen and thinkers.[2]

Hsün Yüeh wrote the *Han-chi* and the *Shen-chien* during this crucial period of transition from the Han to the post-Han era. In a sense, he was the last of the Han Confucian moralists. Many of his dogmatic statements in the *Han-chi* and the *Shen-chien* served as a testimonial to the moral idealism at the core of Han Confucianism as it transmutated through the *ch'ing-i* movement during the last century of the Later Han. What is particularly noteworthy is that, even while reiterating the Han Confucian moral dogmas, Hsün Yüeh was casting serious doubt on the philosophical basis of moral dogmatism. His reflections on the impermanent and ineffable Way, on the limitation of human understanding, and on the problems of morality tended to strike at the very root of orthodox Han Confucian assumptions. In this sense, he is more akin to the post-Han spirit of skepticism which heralded the age of neo-Taoism and Buddhism than to the orthodox thinking of the majority of the Han Confucians. In fact, some of the themes discussed by Hsün Yüeh—such as the primacy of human feelings, the discrepancy between knowledge and truth, and the relationship between moral training (*li-chiao* 禮敎, or *ming-chiao* 名敎 in post-Han parlance) and naturalness (*tzu-jan*)—as well as the vocabulary he used, his subtle manipulation of its dialectical meanings, and even the peculiar laconic style of some of his essays tended to set the pattern for the works of the *ch'ing-t'an* (Pure Conversation) and *hsüan-hsüeh* (Mystical Learning) movement of Wei (A.D. 220–

---

[2] For a survey of the intellectual scene, especially the interchange between the Confucians and the Taoist-Buddhists immediately following the fall of the Later Han, see Erik Zürcher, *The Buddhist Conquest of China* (Leiden, 1959). See also Richard B. Mather's introduction to his translation of the *Shih-shuo hsin-yü, A New Account of Tales of the World* (University of Minnesota Press, 1976), xvi-xxvi; and Donald Holzman, *Poetry and Politics: The Life and Works of Juan Chi, A.D. 210–263* (Cambridge University Press, 1976).

INTRODUCTION

264) and Chin (A.D. 265–420) times.[3] Historically, Hsün Yüeh thus closed the last chapter of Han thought and opened a new phase in China's intellectual history.

○ ○ ○

From a broader perspective, late Han and post-Han skepticism as a reaction to Han dogmatism appears to be an obverse of the positivist reaction in late classical and early Han times against the strains of agnosticism and skepticism in early classical thought. In this respect, Hsün Yüeh's cautious skepticism and some of his un-Confucian views were closer to the pristine Confucian spirit than the transmutated Han orthodoxy had ever been. More importantly, as the following analysis shows, many of Hsün Yüeh's critical discourses and philosophical dialogues, presented in an elliptical style in the present *Shen-chien*, become profoundly meaningful in the context of a dialectic tension between skepticism and dogmatism in ancient Chinese thought as it evolved from the late Chou period to late Han times.

Classic schools of Chinese thought originated in a time of many uncertainties. The breakdown of the Chou order had discredited the clan rules and cultic dogmas of old. Man could no longer rely on the traditions of religion and ritual to provide answers to the questions arising from a rapidly changing world. Even the supreme God and Heaven appeared to be remote and unresponsive to man. It was under these circumstances that Confucius (551–479 B.C.) and a host of other thinkers in late Chou times advanced their secular philosophical teachings.[4] Underneath Confucius' political and

[3] Chi-yun Chen, *Hsün Yüeh*, pp. 129, 136–147, 164–165, 170–171.

[4] Fung Yu-lan, *A History of Chinese Philosophy*, Vol. 1, (Princeton University Press, 1952), pp. 7–54; Wing-tsit Chan, *A Source Book in Chinese Philosophy* (Princeton University Press, 1963), pp. 3–13; H. G. Creel, *Confucius and the Chinese Way* (Harper Torchbook, 1962 reprint), pp. 12–24. A more detailed analysis of the socio-political changes may be found in Cho-yun Hsü, *Ancient China in Transition* (Stanford University Press, 1965); also Kuan Feng and Lin Lü-shih, "The Birth of Materialist Philosophy at the End of the Western Chou and the Beginning of the Eastern Chou," *Chinese Studies in Philosophy* 2:1–2 (1970–1971), 54–79.

INTRODUCTION

ethical pronouncements, there was a deepset uneasiness about man's ability to understand the ultimate meaning of Heaven and the Way. It was from this uneasy agnostic standpoint that Confucius proffered his assurance of pragmatic human efficacy.[5]

Confucius' idea of human efficacy led to the grandiose visions of Mo Tzu (fl. 479–438 B.C.) and Mencius (fl. 371–289 B.C.), who saw an evolution of human unity and order on the basis of "love," justice, and benevolence.[6] But the Mohist and Mencian attempts to justify their teachings by appealing to the "Will of Heaven" or to the cosmic essence of goodness tended to obscure the cautious agnosticism of Confucius. The debates between the Mohist and Mencian schools over the doctrines of their masters led to a new upsurge of skepticism expounded by the Sophists and the Taoists in the middle of the fourth century B.C.[7] The Sophists or School of Names (*Ming-chia*) adhered to a nominalism which asserted that human ideas were merely names and words, whose meanings were derived from the named reality and could not transcend it or be detached from it.[8] The Taoists

---

[5] Fung Yu-lan, *History*, Vol. 1, pp. 57–59; Wing-tsit Chan, *Source*, pp. 14–48; Creel, *Confucius*, pp. 75–99, 109–141.

[6] Fung Yu-lan, *History*, Vol. 1, pp. 91–100, 119–131; Wing-tsit Chan, *Source*, pp. 49–60, 211–226.

[7] A. C. Graham, "The Place of Reason in the Chinese Philosophical Tradition," in Raymond Dawson ed., *The Legacy of China* (Oxford University Press, 1964), pp. 30–45; Frederick W. Mote, *Intellectual Foundations of China* (New York, 1971), pp. 51–60; Seichi Uno, "Some Observations on Ancient Chinese Logic," *Philosophical Studies of Japan* 6 (1965), 31–42.

[8] Fung Yu-lan, *History*, Vol. 1, pp. 192–220. The emphasis of Fung and Bodde on ideational "universals" oriented their analysis of the *Ming-chia* statements on "the white horse," "the hard and white stone," and "meaning (*chih*) and things (*wu*)" in a different direction. As "nominalists," the *Ming-chia* thinkers were troubled by the discrepancy between "name" (*ming*) and "reality" (*shih*); hence their special concern over the relationship between "meaning" (*chih*) and "things" (*wu*). "A white horse is not a horse" because in reality when you call for a white horse, a yellow horse, which is by name a horse, will not do, and because "whiteness" is a reality in the real "thing" (the white horse) and cannot be conjured away by mere names (the categorical horse); *Kung-sun Lung tzu*

INTRODUCTION

viewed reality as essentially relativistic and human reason as ultimately illusory.[9] Although the Sophists and the Taoists thus denigrated Confucius' central idea of human efficacy, their skepticism was more congenial to Confucius' agnostic attitude than were the ideas of many later Confucians.

The agnosticism and skepticism of the early classical thinkers were overwhelmed by the positivist teachings of the Cosmologist Tsou Yen (305–240 B.C.), the Confucian Hsün Tzu (fl. 298–238 B.C.), and the Legalist Han Fei (d. 233 B.C.) in

---

(*SPPY* ed.), 3b–5. For a name (or idea) to be meaningful (*chih*), there must exist things-in-themselves (*wu*, such as horses) to be named; there must be the act of naming (man pointing to, *chih*; or calling for, *wei* 謂) them; and there must be a particular (this and not that) object (the named thing—this white horse) corresponding (*ying* 應 or *wei* 唯) to the name, *ibid*. 5b–6, 11b–12. Cf. A. C. Graham, "Kung-sun Lung's Essay on Meaning and Things," *Journal of Oriental Studies* 2:2 (1955), 282–301, and "Two Dialogues in Kung-sun Lung Tzu," *Asia Major* (N. S.), 11:2 (1965), 128–152; Y. P. Mei, "Some Observations on the Problems of Knowledge among the Ancient Chinese Logicians," *Ts'ing-hua Journal of Chinese Studies* 1 (1956), 114–121; Chung-ying Ch'eng and Richard S. Swain, "Logic and Ontology in the *Chih-wu lun* of Kung-sun Lung Tzu," *Philosophy East and West* 20:2 (April 1970), 137–154.

[9] Joseph Needham, *Science and Civilization in China*, Vol. 2 (Cambridge University Press, 1956), pp. 86–98; Wing-tsit Chan, *Source*, pp. 186–188; A. C. Graham, "Reason," pp. 39–43. The Taoist distrust of man-made categories paralleled the *Ming-chia* distrust of mere names; see Nai-t'ung Ting, "Laotzu's Critique of Language," *Etc., Review of General Semantics* 19:1 (May 1962), 5–38, and "Laotzu: Semanticist and Poet," *Literature East and West* 14:2 (1970), 212–244; Chad D. Hansen, "Ancient Chinese Theories of Language," *Journal of Chinese Philosophy* 2:3 (June 1975), 252–257, 272–282. The debate on the special features of Chinese language, logic, and ways of thinking is ongoing; see Chung-ying Ch'eng's survey article, "Inquiries into Classical Chinese Logic," *Philosophy East and West* 15 (July-October 1965), 195–216; Von Rolf Trauzettel, "Zum Problem der chinesischen Ontologie unter dem Aspekt der Sprache," *Zeitschrin der Deutschen Morgenländischen Gesellschaft* 119:2 (1970), 270–277; Henry Rosemont Jr., "On Representing Abstractions in Archaic Chinese," *Philosophy East and West* 24:1 (January 1970), pp. 71–88; Donald J. Munro, *The Concept of Man in Early China* (Stanford University Press, 1969), pp. 52–58. The language problem in *Kung-sun Lung tzu* is evident in its use of the same word *chih* (originally a pointing finger) to mean the referring agent, the act of reference, and the intended meaning. The Taoist distrust of man-made categories is evident in the very first sentence in *Lao-tzu*.

late classical thought, as well as the Huang-Lao school of Taoism and the eclectic Confucians of early Han times. The breakdown of the old Chou order, which gave rise to naked power struggles and incessant warfare among the contending states, had led many early classical thinkers to question the effability of Heavenly reason and the efficacy of man-made systems. However, as the historical process proved to be irreversible, it established its own validity and its logic of inevitability in the human mind. Politically, the collapse of the Chou authority led to the formation of those powerful Warring States that laid the foundation for the emerging unitary empire. Intellectually, the decline of cultic traditions released the creative secular spirit in man, strengthening his reasoning power and his sense of self-reliance in facing the unknown. "It is man who broadens the Way, not the Way that broadens man"[10]—Confucius' notion of pragmatic human efficacy thus became the foundation stone for a positive anthropocentricity which he had not anticipated.

There are other reasons for this change from agnosticism and skepticism to positivism and dogmatism in late classic Chinese thought. First of all, underlying the agnostic attitude of Confucius and the skepticism of the Sophists and the Taoists, there is a basic anxiety about the uncertainty of the absolutes and a common quest for an alternate assurance. The ineffability of the Way of Heaven prompted Confucius to rely on the Way of man. The limitation of human reason and the impermanence of human existence prompted the Sophists and the Taoists to search for the absolute in nature or the cosmos. Secondly, a heightened awareness of the human predicament often leads to a radical affirmation of human reality, as is the case in classical Chinese thought and in modern European existentialism.[11] If, according to the

[10] James Legge, I, p. 302, Cf. Fung Yu-lan, *History*, Vol. 1, p. 32.

[11] For a recent study emphasizing the traditional Chinese feeling of conflict and predicament, which thus repudiates Max Weber's stereotype of Confucian conformity and complacence, see Thomas A. Metzger, *Escape from Predicament* (Columbia University Press, 1977), pp. 3–14 and *passim*.

INTRODUCTION

Sophists, the universe has no center, then everywhere is its center; according to the Taoists, if everything is relativistic, then relative existence and truth will be the ultimate and hence the only absolutes attainable by man in his particular mode of existence.[12]

The sad fact of agnosticism and skepticism is that it requires a discipline of mind to suspend judgment, to accept the unknown, and to endure the anxiety of the uncertainties of life. It is a discipline attainable by only a few and a state of mind difficult to sustain for long. As the changes brought about by the decline of the old Chou traditions and the rising power of the new states affected a larger population, the demand for practical action and systematic doctrine thus overshadowed the philosophical reservations of the agnostics and skeptics.[13]

Attention was now focused on society, culture, and the state as the concrete, enduring realities of human existence. For, while individual man passes on, society and culture continue and cumulatively build on his achievements.[14] From these emerge the structures and patterns, conventions and rules, ideas and knowledge—*li* and *tao*—which not only testify to the efficacy of human action and reason in the past, but also make action and reason effective in the present.[15] According to Hsün Tzu, it is immaterial that human nature

---

[12] See the ninth paradox of Hui Shih, "I know the center of the world; it is north of the north (i.e., Yen, a northernmost state in China of the Warring States period) and south of the south (i.e., Yüeh, a southernmost state)," *Chuang-tzu* (SPPY ed.), 10:21; Wing-tsit Chan, *Source*, p. 234; cf. Bernard S. Solomon, "The Assumptions of Hui-tzu," *Monumenta Serica* 28:1 (1969), 16–17. For studies of the "positivist" aspect of Taoism, see the discussion and translation of the first two sentences of *Lao-tzu* by J. J. Duyvendak, *Tao-te ching* (London, 1954); H. G. Creel, *What is Taoism?* (University of Chicago Press, 1970), pp. 1–24, 37–47 and *passim*. See also note 36 below.

[13] Henry Rosemont Jr., "State and Society in *Hsün Tzu*," *Monumenta Serica* 29 (1970–1971), 38–78.

[14] According to Hsün Tzu, knowledge and morality are produced by cumulative (*tsi* 積) human efforts; *Hsün-tzu* (SPPY ed.), 1:1–4; 4:12; 17:6b.

[15] Fung Yu-lan, *History*, Vol. 1, pp. 278–299.

INTRODUCTION

might be bad[16] or that human thinking might err;[17] society and culture, through human effort, will rectify man's error and make him good. Hsün Tzu argued that what is bad is an error; it errs in harming and destroying the society and the culture and as a result eliminates itself in the destructive process.[18] What is good, on the other hand, contributes to the preservation and advancement of society and culture and as a result is itself preserved and advanced in the constructive process to become an efficacious social and cultural legacy (*li* and *tao*). Those who had distinguished themselves in this constructive process became the Sage-kings. Their accomplish-

[16] *Hsün-tzu*, 17. Hsün Tzu's thesis that human nature is bad has been emphasized by many scholars in contrast to Mencius' thesis on the goodness of human nature. The idea of man's innate goodness as a cosmic essence and as Heaven's endowment in every man was central to Mencian thought. The idea of man's innate badness only served to bring out the importance of human effort and as such it was of secondary importance in Hsün Tzu's thought, in which even Heaven played a secondary role, 11:9–13. To Hsün Tzu, it is irrelevant that the Sage-kings might have the same innate nature as a wicked ruler or a bandit chief, 17:5. In the same essay, Hsün Tzu mentioned clear eyesight and good hearing as part of man's innate nature, which cannot be considered all evil, 17:2a. Elsewhere, he called man's sense organs Heavenly endowed faculties, 11:10a. The crucial statement in *Hsün-tzu* is that "man's innate nature is inadequate and incapable of ordering itself by itself [for the sake of goodness]," 4:11b. Cf. Donald J. Munro, *Concept*, pp. 77–78. The question of whether human nature is good or bad remained of secondary importance in Han Confucian thought, which was by and large molded by Hsün Tzu's teachings; see discussion below; also *SC* 5.15. According to Yang Hsiung (53 B.C. to A.D. 18), *Fa-yen* (*HWTS* ed.), 1:1b–2a, Wang Ch'ung (A.D. 27–97), *Lun-heng* (*HWTS* ed.), 2:14–16, 3:16–21, and Wang Fu (ca. A.D. 90–165), *Ch'ien-fu lun* (*HWTS* ed.), 1:3, 3:9, man's nature may be good or bad, but the important point is that it can be transformed (*hua* 化) by human effort.

[17] Confucius mentioned that it is man who opens the Way (see note 10). But according to *Hsün-tzu*, it is the Way that regulates man (15:4); the Way embodies the accomplishments of the former Sages (15:7) and man must learn from the wisdom of the former Sages and follow the regulations of former Sage-kings (15:9–10). Names (language and ideas) are the product of culture, molded by conventional usage, and based on commonly agreed meanings (16:1a, 3a, 4b). Cf. Fung Yu-lan, *History*, Vol. 1, pp. 305–309; Donald J. Munro, *Concept*, pp. 77, 81.

[18] Cf. Fung Yu-lan, *History*, Vol. 1, p. 297.

INTRODUCTION

ments were accumulated in the enduring tradition (*li*), and systematized and expounded in the permanent Canons (*ching* 經) by the Sage Confucius, to be transmitted to later generations.[19] Confucian learning and teaching thus became indispensable for the continuity of civilization.

The emphasis on the supremacy of society and human efficacy paved the way for the prevailing order to be regarded as the validation of truth and goodness.[20] According to the Legalist, the state is the product of advanced culture and the most effective organization of society.[21] Both Hsün Tzu and Han Fei argued that as time passed the tradition (*li*) of the former Sages must give way to the law (*fa*) of the later kings.[22] As natural order produced and was superseded by social order, social order in turn produced and was superseded by political order.[23] The power and reason of state thus predominated as the most effective mold of human action and reason in the Legalist doctrine.[24] In 220 B.C., with the unification of China by the Ch'in, Legalism was proclaimed as the sole official ideology of the regime. Human reason thus seemed to triumph —reason in Hsün Tzu's sense of that which was embodied in the progress of human society and culture, and in Han Fei's sense of that which guided the advanced form of state, exemplified by the Ch'in empire. Confidence in man, the optimism that he can open his way and control his destiny, seemed to have been vindicated by history.[25]

[19] *Ibid.*, pp. 294–299.

[20] In *Hsün-tzu*, the Sage-ruler was supposed to set the moral norm and the standard of truth for his people; see notes 17, 18, 19; *Han Fei tzu* (*SPPY* ed.), 2:3a–5, 7, 10–11. Cf. Derk Bodde, *China's First Unifier* (Leiden, 1938), pp. 187–211.

[21] *Han Fei tzu*, 19:1. Cf. A. C. Graham, "Reason," p. 48; Wm. Theodore de Bary et al. ed., *Sources of Chinese Tradition* (Columbia University Press, 1960), pp. 129–130; Wolfgang Bauer, *China and the Search for Happiness* (New York, 1976), pp. 56–61.

[22] *Han Fei tzu*, 19:1–6; 5:7b. Also note 20.

[23] See note 22.

[24] *Han Fei tzu*, 2:4b–8a; 12b–14a; 8:7–8a and *passim*. Derk Bodde, *China*, pp. 41–43, 211–215.

[25] See the court's proclamations in *Shiki*, 6:32, 33–38, 42–44. For the Con-

INTRODUCTION

Emphasis on the efficacy of human reason and experimentation gave rise to another trend in late classic Chinese thought —the anthropocentric cosmology of Tsou Yen. The origin of the *yin-yang* and "Five Elements" cosmological concepts is still an open question. Primitive ideas about the duality of day and night, sun and moon, light and darkness, male and female, about the seasonal changes and the cardinal points, and about the influence of nature and the supernatural on man are perhaps as old as Chinese civilization itself.[26] What distinguishes Tsou Yen's cosmology is his bold attempt to construct a grandiose scheme to correlate history and human affairs with cosmic cycles of change (the mutation of the *yin-yang* and the "Five Elements"). Although Tsou Yen's idea of cosmic and human changes was akin to the Taoist, his confidence in the efficacy of human reason and action was a far cry from Chuang Tzu's skepticism. Like Hsün Tzu and Han Fei, Tsou Yen was a systematizer.[27] It seems no mere accident that his cosmology, alongside Legalism, became dominant at the new imperial court.[28]

---

fucian contribution to such anthropocentric optimism, see Thomas A. Mertzger, *Escape*, p. 264, note 243.

[26] A. Forke, *Lun-heng* (1911; 1962 reprint), Vol. 2, pp. 431–478; Fung Yu-lan, *History*, Vol. 1, pp. 26–30, 32–33. In assigning much later dates (4th and 3rd centuries B.C.) to this tradition, Joseph Needham mentioned but did not give adequate consideration to the distinction between the primitive notions of *yin-yang* and "*Five Elements*" on the one hand and their systematization by Tsou Yen or in the *I-ching* on the other hand; *Science*, Vol. 2, pp. 216, 232, 273–274. The notions of the "four directions" (these and the center make five) and "gods of the four directions" had already appeared in the Shang oracle scripts, Ch'en Meng-chia, *Yin-hsü pu-tz'u tsung-shu* (Peking, 1956), pp. 319–321, 584–594. A strong dualism was suggested by the positive-negative pairs of Shang oracular questions and their possible answers ("yes" or "no"), cf. Chou Hung-hsiang, *Pu-tz'u tui-chen shu-li* (Hong Kong, 1969), as well as by the Shang sacrificial rituals; K. C. Chang, *Early Chinese Civilization* (Harvard University Press, 1976), pp. 84–89, 93–114. Cf. Kuan Feng and Lin Lü-shih, "Thought of the Yin Dynasty and the Western Chou," *Chinese Studies in Philosophy* 2:1–2 (1970–1971), 24–40.

[27] Fung Yu-lan, *History*, Vol. 1, pp. 159–169; Joseph Needham, *Science*, Vol. 2, pp. 232–241.

[28] *Shiki*, 6:23–25; Derk Bodde, *China*, pp. 112–115.

INTRODUCTION

As mentioned above, the positivist teaching of the third century B.C. arose against a basic anxiety over the ultimate meaning and reality of existence. Hsün Tzu's attempt to affirm the validity of human value within the confines of the social process and cultural tradition and Han Fei's effort to restrict it to the operation of the state were justified as long as they limited themselves to the particular sphere of human efficacy. The cosmological speculation of Tsou Yen, however, was a further attempt to assuage such anxiety by a formula purporting to reduce the unknown to the known and thus project human reason far beyond its proper limit. This form of cosmological speculation produced a mechanistic system to calculate the "mutual correspondence between Heaven and Men."[29] It often involved crude interpretation and manipulation of omens to further political ends or to uphold popular superstitions.[30] Philosophically, it led to a serious abuse of the Chinese anthropocentric world-view.

This misplaced confidence in human efficacy reached a climax when the First Emperor of Ch'in considered himself to be not only the greatest of all Sage-rulers but also the pivot of cosmic harmony.[31] The collapse of the Ch'in empire in 206 B.C., barely sixteen years after its inception, thus had profound and far-reaching implications for the subsequent development of Chinese thought. It indicated that human constructs, even though they might be as powerful as the Ch'in empire, could not withstand the onslaught of history; that human reason, even though it might be as sophisticated as Legalism or as bold as the calculations of the cosmologists, could not comprehend the variability of change in history or cosmology. The prag-

---

[29] Joseph Needham, *Science*, Vol. 2, pp. 238, 247–267.
[30] Hans Bielenstein, "An Interpretation of the Portents in the *Ts'ien Han Shu*," *Bulletin of the Museum of Far Eastern Antiquity* 22 (1950), 127–143. Wolfram Eberhard, "The Political Functions of Astronomy and Astronomers in Han China," in John K. Fairbank ed., *Chinese Thought and Institutions* (University of Chicago Press, 1957), pp. 33–70.
[31] See note 25.

matic mind of man failed to anticipate the course of history, but was more effective as its post-mortem judge, as the ruthless denunciation of Ch'in Legalism by later thinkers demonstrated.[32] In the long run, the disaster of Ch'in Legalism tended to undermine all philosophical-ideological systems in the Chinese mind. For if classical thought by its inner dynamics had given rise to Legalism as its culminating philosophical-ideological system, then the discrediting of Legalism reflected on the shortcomings of all ideal systems as such. With the rise of the Han dynasty, a more circumspect eclecticism evolved from a somber mood.[33]

The triumph of Confucianism as the orthodox teaching of the Han empire was not so much the domination of one school of thought, but rather the development of a syncretism that incorporated the teachings of many different schools of classical thought. Han Confucianism was an eclectic tradition, born of a profound wariness of pure ideas and ideological systems and strengthened by a new concern for the lessons of history and the value of education and culture—notions that coincided with the central ideas of Confucius and his followers.

Confucianism did not gain ascendance in the new empire immediately. The early Han court reacted to Legalism by favoring the Taoist teaching of non-action and quietism, moderation and restraint—a way of reckoning with the futility of excessive reason and action. In politics, this meant that the regime made no serious attempt to interfere with ongoing Legalist administrative procedures at lower levels of govern-

[32] *Shiki*, 6:87–103, 110–114; 48:20–25; 97:16. *HSPC* 50:1; 51:1–6, 9a.

[33] According to Fung Yu-lan, *History*, Vol. 2, pp. 2–6, thence ended the age of creative philosophy and began the prolonged period of classical learning in China. Both Fung and H.G. Creel, *Chinese Thought from Confucius to Mao Tse-tung* (Mentor Book, 1953), pp. 132–152, considered the constrictive environment of the unitary empire to be the main reason for the stagnation of Chinese thought in the post-Chou periods. In the following discussion, more attention will be given to the internal dynamics of the development of Han thought, although external political factors will also be considered.

INTRODUCTION

ment or to change the conditions in local communities.³⁴ This allowed the state and society to recuperate of their own accord, but it also enabled Legalism to survive, making the denunciation of Legalism a negation more in name than in deed. On the other hand, these post-Ch'in Legalists could no longer openly profess Legalism nor advocate its theories. Only by demonstrating expertise in government affairs and practical knowledge of statecraft could they continue to function in the new regime. Expertise in government affairs and statecraft (*li-shih* 吏事) henceforth constituted a tradition Legalistic in fact but not in name.³⁵ The Taoist teaching of non-action thus contributed much to the evolving Han eclecticism.³⁶

Other Taoist concepts also deeply affected Han thinkers. They saw the disintegration of the Ch'in as a tragic lesson that substantiated the Taoist view of history, time, and change. Taoist concepts of impermanence, relativity, and cyclical

---

³⁴ The doctrine was represented by Chang Liang, Ts'ao Ts'an, and Ch'en P'ing, all important advisers to Emperor Kao-tsu (r. 202–195 B.C.) in the founding of the dynasty. Ts'ao Ts'an became a prime minister in 193 B.C., succeeded by Ch'en P'ing in 189 B.C.; *Shiki*, 54:13–17; 55:4–7, 28–30; 56:20, 23; Burton Watson tr., *Records of the Grand Historian of China* (Columbia University Press, 1961), Vol. 1, pp. 135–136, 149–150, 164–165, 167, 421–425. The principle became firmly entrenched from the time of Emperor Wen (r. 179–157 B.C.) until the early years of Emperor Wu (r. 140–87 B.C.); *Shiki*, 10:38–43, 47; 12:3; 107:8–9; 121:6–7; *HSPC* 3:8a; 4:19–21a; 5:10b; 37:4b; 46:8a; 50:9, 15a.

³⁵ Many high officials at the early Han court had served in lower government positions under the Ch'in dynasty; *Shiki* 121:6; *HSPC* 2:1; 4:1a; 12:1, 2a, 3b, 4b; 13:1a; 39:1, 7a, 13a. Cf. Ch'ien Mu, *Ch'in Han shih* (Hong Kong, 1957), pp. 41–46, 66–69; Dubs, II, pp. 20–22. The situation was pointed out by Chia I in his memorials to Emperor Wen, *HSPC* 48:18b–21a, 25, 27b–31; and by Lu Wen-shu, *HSPC* 51:31b–33a. Both Chia I and Lu Wen-shu were themselves influenced by Legalism; *HSPC* 48:13b; 51:30b. See also Wang Ch'ung's discussion of the *wen-li* in note 82 below.

³⁶ This is especially true of the Huang-Lao school of Taoism in early Han times. Large collections of writings related to this tradition have recently been unearthed from Han tombs; cf. Jan Yün-hua, "The Silk Manuscripts on Taoism," *T'oung-pao* 63:1 (1977), 65–84; also Michael Loewe, "Manuscripts Found Recently in China: A Preliminary Survey," *T'oung-pao* 63:2–3 (1977), 118–120.

INTRODUCTION

reversal in relation to personal and political success and failure gained prominence, as did the focus on the importance of that which was timeless.[37] The study of change and continuity in history—learning the lessons and preserving the values of the past—became the province of the early Han Confucians. The debacle of the Ch'in, the Confucians argued, was due to the denigration of these lessons and values by the Ch'in Legalist ruler.[38]

These early Han Confucians were able to persuade the regime to introduce a number of reforms—changes which seemed quite modest at the beginning, but eventually had an enormous impact on subsequent developments in Chinese thought and culture. The early Han court nullified the Ch'in decrees for the suppression of learning and proscription of books.[39] Earnest efforts ensued to revive the various schools of thought, to recover ancient books and historical documents, and to preserve contemporary records.[40] The Confucians expanded the scope of their historical studies beyond the Ch'in dynasty to encompass the evolution of the Warring States, the moral lessons in the *Spring-and-Autumn Annals* (*Ch'un-ch'iu*), and the idealized traditions of the "Three Ancient Dynasties" of Hsia, Shang, and Chou and of other Sage-rulers of antiquity. They analyzed history in terms of strategic, pragmatic, moralistic, and cosmological considerations.

In ca. 176 B.C. Emperor Wen established the precedent of appointing scholar-tutors to the imperial princes.[41] In 136 B.C.

---

[37] Important examples are the "Appendixes" to the *Book of Changes* and the *Huai-nan tzu*. Fung Yu-lan, *History*, Vol. 1, pp. 379–399; Wolgang Bauer, *China*, pp. 71–76.

[38] *Shiki* 6:86–103; *HSPC* 31:24b–29; 43:7; 48:18b–20a, 21–27.

[39] *HSPC* 2:5a, 30:1; *Shiki* 121:6.

[40] Ch'ien Mu, *Ch'in Han shih*, pp. 68–69; Dubs, I, pp. 216–217. The Five Confucian Classics (*ching*) were collections of heterogeneous ancient documents considered to be valuable because they preserved the traditions of the ancient Sage-rulers. Works of the Confucian masters such as *Mencius* and *Hsün-tzu* on the other hand were classified as philosophical works (*tzu*), together with the works of Taoism, Legalism, etc. *HSPC* 30:2b–52a; Ch'ien Mu, *Ch'in Han shih*, pp. 81–84.

[41] *HSPC* 48:2a, 9a; 49:9a; Dubs, I, pp. 216–217.

INTRODUCTION

Emperor Wu (r. 140–87 B.C.) designated a corpus of ancient writings known as the Five Classics (*ching*) as the curriculum of the Official Erudites (*po-shih*), and twelve years later a group of official disciples and students were assigned to study under the guidance of such Official Erudites in preparation for governmental appointments. The number of disciples increased from 50 to 3,000 by the end of the Former Han dynasty. Meanwhile, similar programs were begun at the commandery and district levels. By the Later Han, students at the Imperial Academy (*T'ai-hsüeh*) exceeded 30,000, which probably included many unofficial students clustered around that symbol of scholarly prerogative.[42]

Out of such government-sponsored educational programs arose an imperial ideology. Although it was essentially Confucian, the new orthodoxy did not prescribe the *Analects* of Confucius, *The Work of Mencius*, or the writings of Hsün Tzu as part of the official curriculum. Instead, the Five Classics consisted of a body of works considered to have profound historical significance and permanent cultural value. While they were attributed to Confucius or were at least regarded as products of the Confucian school, these writings actually had quite heterogeneous origins and divergent implications. *The Book of Changes* (*I-ching*) and "The Doctrine of the Mean" (*Chung-yung*, a treatise in the *Records of Rites, Li-chi*) were as much Taoistic as Confucian; *The Book of Historical Documents* (*Shu-ching*) and the *Spring-and-Autumn Annals* (*Ch'un-ch'iu*) were susceptible to strongly Legalistic interpretations.[43]

Adding to the diversity of the Confucian orthodoxy was the fact that scholars trained in various schools of classic thought were recruited to the government schools. Thus a student might devote the prime of his life to the study of Taoism or Legalism, then enter the Imperial Academy and spend two or three years mastering one of the Five Classics, pass the required examina-

[42] *HSPC* 88:3b–6a; Dubs, II, pp. 20–25; *CC* 79a (*lieh-chuan* 69A): 2b; 67 (*lieh-chuan* 57): 3a; P'i Hsi-jui, *Ching-hsüeh li-shih* (Peking, 1959), p. 101; Ch'ien Mu, *Ch'in Han shih*, pp. 187–209.
[43] Fung Yu-lan, *History*, Vol. 1, pp. 400–407. Tjan Tjoe Som, *Po Hu T'ung* (Leiden, 1949), pp. 95–100.

17

tion and receive an appointment in government—and henceforth use his Taoistic or Legalistic interpretation of the Classics to justify his official conduct.[44] What was emphasized in the orthodoxy was not philosophical or doctrinal purity but an immersion in historical traditions and a reconciliation of divergent human values. It was an attitude that outlasted the vicissitudes of all formal schools of thought and religion in traditional China.

Han Confucianism was, however, more than an attitude or an awareness. As an historical movement, it had a cluster of traditions and basic assumptions, and the movement was embedded in its particular time and institutional context. During the Former Han dynasty, Confucianism was mainly a movement to revitalize and reform the imperial system after the Ch'in failure. The downfall of the Ch'in had been a severe blow to the optimism about the efficacy of human reason and action. The attitude of the early Han Confucians toward the resurging imperial order was quite ambivalent.[45] The Ch'in experience revealed both the limitation and the magnitude of the imperial state as a human construct—a reality that never failed to impress the Chinese mind in subsequent millenia. As peace and order were restored under the Han dynasty, there was a revival of optimism about the efficacy of human endeavor and the function of the state. But the Han Confucians did not place their unmitigated faith in state power. There was the awareness that states and dynasties rose and fell and rulers would err; that only history and culture continued, ensuring permanence and rectifying human folly. The Confucians saw state laws not as an analogue of the laws of nature, objective and absolute, but rather as man-made, catering to particular human needs and subject to the guidance of superior human wisdom as it crystallized in the moral standards and cumulative traditions of society as a whole.

[44] For example, see *HSPC* 49:8; 51:1a, 30b; 52:3b–4; 58:1b, 4b, 11; 59:1–2; 64:14a, 16b; 64B:4a; 71:5b–6a; 75:1a; 76:19b; 84:1; 85:1a; 87A:2; 87B:21; 89:4.

[45] Dubs, I, pp. 15–27; *Shiki* 99:2–5, 14–15; Burton Watson tr., *Records*, I, pp. 285–287, 293–294.

INTRODUCTION

Early Han Confucian thought thus reverted to the basic stand of Hsün Tzu. However, having observed the failure of Legalism and the fall of the mighty empire of Ch'in, the Confucians were less sanguine about man's practical reason. Hsün Tzu might be correct in trusting that the collective wisdom of society and the cumulative weight of culture would rectify human errors and sustain the Way. But the price to be paid for such gigantic errors as the Ch'in debacle was too high to be entrusted to unknown destiny or to historical hindsight. And, among individual men, who was to muster the wisdom of society and the Way other than the ruler? But the ruler could also be horribly mistaken, like the emperor of the Ch'in, or an insurgent like the founder of the Han who had to triumph through the hell-fire of revolution. No, there ought to be a better way. The old Chou concept of Heaven (*t'ien*) and its Mandate (*ming*) was revitalized to become the underpinning of the Han Confucian world-view.[46]

Although Confucius had questioned man's ability to understand Heaven, there was some indication that he retained a strong personal faith in divine Heaven and a feeling of sacred mission in his work among men.[47] He believed that by knowing and serving humanity one was realizing at least part of Heaven's work and thus fulfilling Heaven's destination (*ming*) for men. The same reverence was expressed by Mencius, who exhorted, "He who has exhausted his heart and mind knows his nature; knowing his nature, he knows Heaven."[48] Inherent in these ideas is that man's knowledge of Heaven is not automatic, nor is his action in fulfilling the will of Heaven. Man has to be trained for this.[49] Education is necessary—either in Confucius'

[46] Wolfgang Bauer, *China*, pp. 71–77; H. G. Creel, *The Origins of Statecraft in China* (University of Chicago Press, 1970), pp. 44–45, 82–85, 93–99, 493–505.

[47] Fung Yu-lan, *History*, Vol. 1, pp. 57–58.

[48] Legge, II, pp. 448–449. For a more enthusiastic exposition of this view by the neo-Confucians and some modern Chinese scholars, see Thomas A. Metzger, *Escape*, pp. 32–39, 118–122 and *passim*.

[49] Cf. Legge, I, pp. 146–147, 186, 190, 250, 257, 258, 313–314, 318; II, pp. 167–168, 188–193, 308–309, 325, 333–335, 363, 372, 404–409, 414, 446–448, 450–451, 489–490, 495–496, 497.

INTRODUCTION

sense of learning to be human (*jen*), Mencius' sense of realizing one's nature, or Hsün Tzu's sense of culture and cultivation. Similarly, implicit in the old Chou notion of the Mandate of Heaven as reiterated by Mencius is that the receipt of such a Mandate by a ruler is not automatic. Although the founding of a dynasty is accorded by the Mandate of Heaven, the succeeding ruler must prove himself worthy of it or the Mandate is lost.[50]

As men learn to improve themselves, to be both good and effective, they elevate not only themselves but also society and culture to a higher level of merit. The person proved to be the worthiest of all would receive the Grand Mandate (*ta-ming*) to become ruler of the state, a political and social hierarchy based on merit. Eventually a perfect system will emerge when every individual has fully realized his human potential through education and has occupied his proper place in the state during the Age of Great Unity (*ta-t'ung*), Ultimate Harmony (*t'ai-ho*), and Universal Peace and Equality (*t'ai-p'ing*). Herein lies the Confucian assumption of the unity of the moral, socio-political, and cosmic orders.[51]

This Confucian assumption was in basic contradiction to the reality of dynastic rule whereby the emperor claimed the highest position in the realm as his hereditary right. This generated considerable uneasiness, which led the early Han Confucians to question whether the infamous Ch'in regime had ever received the Mandate of Heaven and whether the Han dynasty, which triumphed in the bloodletting civil war, was founded in accordance with the Mandate.[52] A saying of about 196 B.C. was: "The

[50] Cf. Legge, II, pp. 292–294, 296–298, 354–361, for Mencius' view. For Han Confucian concepts, see *HSPC* 36:9b–14a, 19b–20a; 56:3a, 4.

[51] Wolfgang Bauer, *China*, pp. 78–85; Donald J. Munro, *Concept*, pp. 112–116.

[52] See note 45. This involved the calculation of the dominant "element" and "virtue" of the Ch'in and the Han dynasties according to the "Five Elements" theory. It was said that the Chou dynasty was of the "Fire element and virtue"; "Water" overcomes "Fire" and hence the dynasty that succeeded the Chou should be of the "Water element and virtue." The Ch'in dynasty thus officially adopted "Water" as its "element and virtue." Chang Ts'ang, the official cosmologist at the early Han court, considered "Water" as still dominant in the Han dynasty. *Shiki* 96:10–12, 17; Watson I, pp. 263–264.

20

INTRODUCTION

Ch'in lost its fortune (*lu* 鹿, deer; a homonym of *lu* 祿, official prerogative). All under Heaven chased after it. He who rode on high horseback or was swift of foot caught it [and became the new ruler]."[53] This uneasiness reinforced the Taoist caution against excessive state power and the Confucian counsel of education and learning. It also revived strains of cosmological speculation which seemed to have suffered a setback from the downfall of the Ch'in. Where the cosmological system developed by Tsou Yen and supported by the Ch'in court implied excessive confidence in human reason, early Han cosmological thought was characterized by a sense of uncertainty, suspended judgment, and a feeling of urgency.[54]

This implied that either the Han merely continued the virtue of the Ch'in and did not effect a change of the "element" or Mandate; or that, if "Water" was the true "element" newly acquired by the Han, directly replacing the "Fire element" of the Chou dynasty, the Ch'in would have been without a true "element" or Mandate but would have merely existed as a transition between the Chou and the Han. The first view accords with the fact that the early Han court made little change of the government system, law codes, and court rituals of the Ch'in. The second view may be derived from the early Han denunciation of the Ch'in (see note 32 above). Cf. Ch'ien Mu, *Ch'in Han shih*, pp. 43–48, 110–113.

[53] *Shiki* 92:40; tr. by Watson, I, 231.

[54] Among the high officials at the early Han court, only Chang Ts'ang was versed in cosmological speculation. His theory was mainly a holdover from the Ch'in; see note 52. It was not until Emperor Wen's reign that Chia I and Kung-sun Ch'en began to propose a radical change of the Ch'in scheme and suggest a new "element of Earth" for the Han; *Shiki* 84:21–22; 10:32–34. Their proposals were opposed and denounced by the Taoist-inclined officials at the court. It was in Emperor Wu's reign that the proposal was adopted in earnest; cf. Michael Loewe, *Crisis and Conflict in Han China* (London, 1974), pp. 17–19, 28–32. Still the endeavor was described unfavorably by Ssu-ma Ch'ien, *Shiki* 6:23–24; 26:9–15; 28:3, 19–27, 35–89. The official stand of the Han court on this issue remained unsettled. Liu Hsiang (77–6 B.C.) eventually expounded the new theory of "Five Elements generating one another" (*Wu-hsing hsiang-sheng*) to replace the old theory of "Five Elements overcoming one another" (*Wu-hsing hsiang-k'e*). He completely revised the scheme of elements ascribed to the ancient dynasties, assigning "Wood" to the Chou, which generated the "Fire" element of the Han, thus eliminating the Ch'in from the scheme; *HSPC* 21B:45–48, 53b, 72.

INTRODUCTION

Cosmology was used by both the Ch'in and later the Han emperors to legitimize their political power. It affirmed the important role played by the emperor in the socio-political, moral, and cosmic spheres in accordance with the Mandate of Heaven. But Han cosmology carried a strong negative undertone. While it purported to study both the cosmic reasons for the rise and fall of earthly regimes and the beneficial or harmful impact of human actions upon the cosmos, and vice versa, the Han Confucians, in incorporating cosmological speculation into their teachings, tended to emphasize mainly its distressing aspects.[55] The early Han Confucians were concerned with the disastrous fall of the Ch'in and the destructive insurrection that gave rise to the Han dynasty. It appeared that whatever the moral worth of the early Han rulers, the founding of the Han dynasty was justified insofar as the destruction of the Ch'in was appointed by Heaven.[56] To the early Han Confucians, the Han dynasty was tenured by an interim grace. It behooved the new ruler to avoid the pitfalls of the Ch'in.[57] Such pitfalls were viewed not only in terms of earthly policies and politics, or long-range history, but, given the current distrust of human rationality, the true reason must be sought from the hidden hand of providence when and where it was revealed to men. Han cosmology came to be known as the study of "disasters and abnormal phenomena" (*tsai-i* 災異)—cosmic distress caused by

---

[55] Han Confucian cosmological theories, especially those developed by Tung Chung-shu (ca. 174–104 B.C.) and Liu Hsiang as preserved in the "Treatise on Astronomy" and "Treatise on the Five Elements" in the *Han-shu*, dealt almost exclusively with abnormal phenomena and disastrous occurrences and their correspondence with man's conduct; *HSPC* 26 and 27A–E. See also Tung Chung-shu's memorials in *ibid.* 56:3a and Liu Hsiang's in 36:9–17, 28–30a. Cf. P'i Hsi-jui, *Ching-hsüeh li-shih*, p. 106; Ch'ien Mu, *Ch'in Han shih*, pp. 209–214.

[56] According to Ssu-ma Ch'ien, the destruction of the Ch'in was mandated by Heaven, but the failure of Hsiang Yü (the most powerful leader in the anti-Ch'in uprising) to found an enduring dynasty was due to some weakness in his character, not to the workings of Heaven; *Shiki* 7:76–77; see also 43:5b, 10b–11, 15a. Cf. Chi-yun Chen, *Hsün Yüeh*, pp. 50–51.

[57] *Shiki* 8:88; 37:13; 55:9. See also notes 32 and 45.

INTRODUCTION

the misconduct of men (especially the ruler) or serving as a divine warning to men for their misdeeds (especially the grave mistakes of the ruler).[58]

As the Han dynasty continued, it seemed to validate its own legitimacy beyond the interim grace.[59] By the time of Emperor Wu, the dynasty reached the apex of its power and glory, and the Confucians were pleased by the success of their reform of learning and education, which established Confucianism as the nominal orthodoxy of the imperium.[60] The concern was no longer whether the Han dynasty had received the Mandate of Heaven or whether the Confucian doctrine would prevail, but when and how the Age of Universal Peace and Equality would arrive.[61] Encouraged by the early success of their reforms, which convinced them of the correctness of their assumptions, the Confucians became more ambitious and dogmatic. Many of them grew impatient with the modest reforms and felt they were in a position to realize their loftier ideal—an Age of Universal Equality or, at least, a state of universal meritocracy.[62]

---

[58] *HSPC* 36:8a; 56:19b–20a; 75:6b–11a; 85:1b, 4b–5a, 7a. See also note 55. Cf. Tjan Tjoe Som, *Po Hu T'ung*, pp. 120–124.

[59] See Ssu-ma Ch'ien's eulogy, *Shiki* 8:88, tr. by Watson, I, pp. 118–119. Compare Emperor Wen's apologetic edicts of 178 and 163 B.C. admitting that he did not deserve the Mandate (*Shiki* 10:20, 35–36, tr. by Watson, I, pp. 351–352, 360–361) and Emperor Wu's edicts of 140 B.C. with respect to the Mandate of Heaven and the attainment of Universal Peace (*T'ai-p'ing*), *HSPC* 56:1b–3a, 8b–9, 13b–14a.

[60] See the historian's eulogy to Emperor Wu, *HSPC* 6:39, tr. by Dubs, II, pp. 119–120; also cf. pp. 20–25, 341–348.

[61] See Emperor Wu's questions and Tung Chung-shu's replies concerning the attainment of Universal Peace, *HSPC* 56:1b–19a. According to Tung Chung-shu, the Han dynasty had achieved worldly power and glory but needed to adopt some fundamental reform measures in order to bring about Universal Peace.

[62] Michael Loewe came to the same conclusion with regard to the triumph of a "Reformist" attitude over the "Modernist" (conservative, Legalistic, political realist?) tradition in the last seventy years of the Former Han dynasty. Loewe's study was based on a different set of events and hence produced a different view of the "Reformist" movement; *Crisis*, pp. 11–13, 93–95, 98–100,

INTRODUCTION

By the middle of the first century B.C., the Confucian ideal of a social and political system based on merit was realized in part under the aegis of the dynastic ruler. Most of the high officials and many middle- and lower-ranking ones were educated and selected according to the Confucian principle of learning and merit. The emperor too received instruction from prominent Confucian tutors before and after ascending to the throne, although he was not selected for the throne by merit but claimed his position by the right of birth.[63] The Confucians thus suspected that the Age of Universal Peace and Equality did not materialize because the ruling dynasty had not been thoroughly reformed or otherwise replaced. As early as 78 B.C. Kuei Hung had memorialized the throne, saying:

"My late master Tung Chung-shu (a prominent Confucian) had said: 'Even though a worthy emperor may have succeeded to the throne and upheld the tradition of culture, this should not prevent a true sage from receiving the Mandate of Heaven to change the dynasty.... Now the Emperor of the Han dynasty should send out messengers to travel across the realm in search of one who is truly worthy, to whom the emperor should yield the throne... in order to accord with the Mandate of Heaven."[64]

Similar suggestions were made by Kai K'uan-jao in 60 B.C. and later by Liu Hsiang (77–6 B.C.).[65] Ku Yung's counsel in 12 B.C. was even more high-sounding. He declared:

"Heaven gives life to the multitude of people, who cannot govern themselves. Therefore Heaven establishes a ruler to administer and manage their affairs. The empire exists for the sake not of the ruler but of the people. Heaven will get rid of a dynasty which loses the Way and let the one in possession

---

106–111, 158–160, 162, 164–165. See my review of this book in the *Journal of Asian Studies* 35:1 (1975), 139–140.

[63] Dubs, II, pp. 196–198, 285–286, 348–353; Ch'ien Mu, *Ch'in Han shih*, pp. 187–193.

[64] *HSPC* 75:1b–2a.

[65] *HSPC* 36:19b–20a, 28b–29; 77:3b–4a. Cf. Chi-yun Chen, *Hsün Yüeh*, pp. 50–51.

INTRODUCTION

of true virtue establish a new regime, thus showing no private favor to one particular house. This makes clear that the realm belongs to all the people in the realm and not to one particular person."[66]

Other Confucians were less sanguine and emphasized the reform of the imperial family by restricting its power, reducing its extravagance, and rectifying its errors.[67]

The fervor for reform reached a climax when Wang Mang, who considered himself a Sage and was praised as such by many of his contemporaries, terminated the Former Han dynasty by forcing its last ruler to abdicate. Wang Mang ascended the throne and founded his new dynasty in A.D. 9, thus fulfilling the Confucian expectation of a Sage becoming emperor. He initiated a series of idealistic reforms. According to Han Confucian doctrines, this should have heralded the Age of Universal Peace and Equality.[68] The dogmatic manner in which Wang Mang handled the problems created by his impractical reforms, and a sequence of natural disasters quite beyond his control, led to the fall of the regime in A.D. 23, barely fourteen years after its founding.[69] To the Han Confucians, the experience was more traumatic than the Ch'in debacle. It cast doubt on the cluster of Confucian ideas about the Sage-ruler, the Mandate of Heaven, the Age of Universal Peace and Equality, and the unity of the socio-political, moral, and cosmic orders.

After the anti-Wang Mang rebellion and turmoil, the Han dynasty was restored (henceforth known as the Later Han) by

---

[66] *HSPC* 85:15a. Cf. Ch'ien Mu, *Ch'in Han shih*, pp. 214–218.

[67] Ch'ien Mu, *Ch'in Han shih*, pp. 218–220; Michael Loewe, *Crisis*, pp. 159–160, 269–270. The motive of these critics was not purely idealistic. Liu Hsiang was embittered by his eunuch opponents at the court; Kai K'uan-jao was prompted by personal frustration; and Ku Yung was a supporter of Wang Mang.

[68] A strong argument for this view was presented by Ch'ien Mu in *Ch'in Han shih*, pp. 269–291. See also Chao I, *Nien-erh shih cha-chi* (Shang-wu yin-shu kuan, 1958), p. 64.

[69] Ch'ien Mu, *Ch'in Han shih*, pp. 282–291; Hans Bielenstein, "The Restoration of the Han Dynasty (I)," *Bulletin of the Museum of Far Eastern Antiquity* 26 (1954), 82–165; Chi-yun Chen, *Hsün Yüeh*, pp. 11–12, 13–15 and notes.

INTRODUCTION

Liu Hsiu (Emperor Kuang-wu, r. A.D. 25–57), a member of a lesser branch of the Former Han ruling house. The reality of empire and dynastic rule thus continued, where human ideals had floundered. Faith in the efficacy of human reason, which had risen to a new dogmatic height after the disillusionment following the fall of the Ch'in, was gravely shaken. History not only defied human anticipation but, it seemed, its purpose and meaning were beyond rational comprehension.

The Han Confucians were in disarray. Many of the privileged (court aristocrats, powerful families, and big landlords) who called themselves Confucian but found their entrenched interests threatened by Wang Mang's reforms were glad that the old dynasty was restored by a new ruler rising from their peers. But there were not a few Confucians whose convictions were as idealistic and dogmatic as those of Wang Mang, whose regime they supported to the bitter end. There were also those who did not actively support Wang Mang's regime, but were nonetheless disillusioned by his downfall or otherwise confounded by the turn of events. The Emperor Kuang-wu, himself arising from the rank of the privileged Confucians, seemed to understand the situation. While rewarding his supporters and strengthening the entrenched Confucian orthodoxy at the court, he also showed his magnanimity in forgiving those who had supported the regime of Wang Mang and in tolerating those who withheld their allegiance to the Later Han on the ground of Confucian moral integrity.[70]

In Later Han times, cosmology fell into disfavor for both political and genuine intellectual reasons. During the declining years of the Former Han, the negative undertone in Han cosmological thought was exploited by the Confucian critics of the throne to further their demand for reform or even for change of the dynasty. It became a dangerous weapon in the hands of Wang Mang, forcing the abdication of the last Former Han ruler. Given the rising Confucian apathy toward the restored

[70] Chi-yun Chen, *Hsün Yüeh*, pp. 11–18 and notes; Ku Yen-wu, *Jih-chih lu* (*Wan-yu wen-k'u* ed.), v. p. 39.

Han dynasty, the court found it unwise to base the legitimacy of the Later Han rule on cosmological speculations or on the rational-moral interpretation of the Mandate of Heaven. The Confucians themselves were confounded by their miscalculation of cosmology concerning the rise and fall of Wang Mang. They doubted whether the cosmos or the fate of men were amenable to human reason.[71]

In Wang Mang's time, Yang Hsiung (53 B.C. to A.D. 18) had suggested that, while a man's practical wisdom might save him from many failings which he erroneously attributed to fate, the true working of fate was determined by Heaven, which was ineffable to man's petty practical mind. Only a great Sage, possessing uniquely superlative wisdom (*tu-chih* 獨智) and a spiritual mind (*shen-hsin* 神心), could comprehend the mysterious workings of Heaven and the cosmos.[72] Pan Piao (A.D. 3–54) contended that the Mandate of Heaven which sanctioned the rise of a dynasty belonged to the realm of the supernatural, which could neither be manipulated nor fully understood by mundane men. Even omens, "disasters and abnormal phenomena," could not be rationally interpreted on the basis of cosmology, but should be taken as mysterious signs of revelation.[73] Alongside the orthodox Confucian moral and cosmological teachings, there arose the apocryphal and prognostic

[71] Su Ching, who had been dean of the Confucian Erudites at the court of Emperor P'ing (r. A.D. 1–5) of the Former Han and remained in high office in Emperor Kuang-wu's reign, mentioned that "many vulgar Confucians of his time were confounded by what they saw and heard [concerning the change of dynasties] and wondered that with the frequent change of dynasties, who was to know which regime was the right one, or that a Sage-ruler might not have been chosen [by Heaven]." *CC* 30A (*lieh-chuan* 20A):1a, 2a; see also *HSPC* 100A:7a–8.

[72] Yang Hsiung, *Fa-yen*, 2:6a; 4:1; 5:2. Yang Hsiung doubted whether an ordinary person was capable of recognizing a great Sage or distinguishing a Sage from a great imposter, 4:4a. He gave his opinion that the Way (*tao*) is like a road or a river (i.e., amoral), upon which all kinds of vehicles or boats travel—the good mixing with the bad, 3:1.

[73] *HSPC* 100A:8–11. Su Ching (see note 71) also had to rely on mysterious signs (*t'u-ch'an*) to support his view that the Han imperial family still retained the Mandate of Heaven; *CC* 30A:2–4.

INTRODUCTION

traditions known as *ch'an* and *wei*, which their supporters claimed to be the esoteric teaching of Confucius, superceding his exoteric doctrines found in the official orthodoxy. This apocryphal and prognostic teaching was strongly favored by the early Later Han court, which placed the fate of a dynasty in the hands of providence, beyond the rational Confucian sanction.[74]

Among Later Han Confucians, some like Huan T'an (43 B.C. to A.D. 28) courageously denounced the apocryphal and prognostic works as superstitious nonsense.[75] But Huan T'an found it imperative to confine his rational discourses within the limit of man's practical common sense. He argued that if a Sage were born again in later times one could recognize his great ability but one would be at a loss to judge whether or not he was a true Sage.[76] Discouraged from their search for cosmological truth and all-encompassing reforms, many later Han Confucians turned their attention to the practical problems of government, reviving the Legalistic concern for statecraft (*li-shih*).[77] Others used their critical abilities for the study of the Confucian classics and history—promoting the unofficial Confucian "Ancient Texts" (*ku-wen*) alongside the official "Modern Texts" (*chin-wen*) of the orthodoxy, collating one with the other and explicating the meanings of both by cross-reference—for the sake of purely academic scholarship.[78]

---

[74] Chi-yun Chen, *Hsün Yüeh*, pp. 15–16; Chao I, *Nien-erh shih cha-chi*, pp. 76–78; Tjan Tjoe Som, *Po Hu T'ung*, pp. 100–128. See Hsün Yüeh's comment, *SC* 3.15.

[75] *CC* 28A (*lieh-chuan* 18A): 3b–5a; Timotheous Polora tr., *Hsin-lin (New Treatise) and Other Writings by Huan T'an (43 B.C. to A.D. 28)* (Michigan Paper in Chinese Studies, 1975), pp. xii–xiii, 65, 238–240 and notes.

[76] "Ch'uan Hou-Han wen," in Yen K'o-chün ed., *Ch'uan shang-ku san-tai Ch'in Han san-kuo liu-ch'ao wen* (Shih-chieh shu-chu ed.), 12:9a; 13:2, 6a–7; 15:7b–8a; Timotheous Pokora tr., *Hsin-lun*, pp. xi, xviii–xx, 10, 25–27, 31, 39–40, 50–51.

[77] See for example Timotheous Pokora tr., *Hsin-lun*, pp. 234–238, 241. A more detailed analysis is included in my chapter on "Crosscurrents of Confucianism, Taoism, and Neo-Legalism in Late Han China" (tentative title) to be published in the *Cambridge History of China*, Vol. I.

[78] P'i Hsi-jui, *Ching-hsüeh li-shih*, pp. 126–127, 142, 148 and notes. See *SC* 2.18.

INTRODUCTION

There were those who lost their interest in both politics and scholastics. They gave up the search for order and meaning in the outside world and turned their attention inward.[79] In early Chinese thought there had been a deep-set polarity between the concern for the "inner realm" and the concern for the "outer realm" of meaning and reality, or between the emphasis on "self-cultivation" and the emphasis on "ordering and harmonizing the world."[80] In late classical thought, the emphasis on the "outer realm" and the effort to reform the world tended to outweigh the other tendency. With the traumatic experience of the Legalistic reforms under the Ch'in and the failure of the Confucian reforms of Wang Mang, Later Han thinkers rediscovered the meaning and appeal of the inner world. Wang Ch'ung (A.D. 27–97) opined that the "fate" of an individual man or of the society and state as a whole was determined by the workings of nature or the chance of events, over which men, even the Sage-ruler, had no control. This included man's "inborn" moral and intellectual qualities as well as his human, natural, and cosmic environments, which were all heteronomously determined and thus extrinsic to man. To Wang Ch'ung, the outside world was made up of interacting fragments; whatever order one might find in this world was but a flimsy mosaic formed accidently by compatible fragments.[81] What was left to the control of the individual man in this world, according to Wang Ch'ung, was his moral intention, determination, and effort. These, he affirmed, were for each man to decide and control, and hence were the only elements intrinsic to him.[82]

---

[79] Chi-yun Chen, *Hsün Yüeh*, pp. 17–18 and notes.

[80] Benjamin Schwartz, "Some Polarities in Confucian Thought," in David S. Nivison and Arthur F. Wright ed., *Confucianism in Action* (Stanford University Press, 1959), pp. 52–58.

[81] Wang Ch'ung, *Lun-heng*, 1:1–17a; 2:4b–9 and *passim*. According to Wang Ch'ung, not only man's success or failure but also his physical condition, moral capacity, and intelligence are all determined by nature, Heaven, or fate. Yet human life or the myriad things are not created intentionally by nature ("Heaven and Earth") but are the result of "accidents"; 3:21b–25.

[82] According to Wang Ch'ung, although the good and the bad in human nature (*hsing*) are predetermined, the bad can be changed into good by human

INTRODUCTION

According to Wang Ch'ung, one might be born of low intelligence and poor health and suffer many reverses in one's social and political life but still be capable of moral goodness. Wang Ch'ung not only criticized Han Confucian cosmology as superstitious nonsense but also revealed the fallacy of the Confucian assumption of the unity of the moral, socio-political, and cosmic orders.[83] Other Later Han Confucians, while agreeing with Wang Ch'ung about the superiority of the inner moral worth of a man in sustaining his humanity and accomplishing sagehood, did not give up hope for an ultimate unity of the "inner" and the "outer" realms. If many Former Han Confucians had made the mistake of looking "outward" for a cosmological order to rectify the socio-political system and mold human morality, the Later Han Confucians—by concentrating on the moral and spiritual strength in man—hoped to extend it outward to influence society and the state and to affect cosmic harmony. The Confucians in the second century A.D., alarmed by the corruption of the society and the state and the breakdown of the Confucian ideological synthesis, made a last effort to salvage them.[84]

According to Wang Fu (ca. A.D. 90–165), the social and political practices of the time were utterly corrupt, and the ruler was far from being a sage.[85] But the potential conflict

---

effort. He declared that moral goodness depends not on human nature but on transforming (*hua*) it (*ibid* 2:14–15). And this transforming effort is decided not by nature or fate but by man's intention or moral will (2:16b), although owing to the haphazard conditions of the outside world "a person rarely fulfills the moral will (*chih*) inside his chest" (2:9b). A person of inner moral strength thus often finds himself in conflict with the outside world (1:6–9). But it is this moral strength and integrity that distinguishes the true worth of the Confucian literati from the earthly power and accomplishments of the "Legalistic officials" (*wen-li* 文吏, see note 35 above), *ibid* 12:1–13:7. The distinction between what is heteronomously determined by Heaven and what is for man to decide was emphasized by Wang Fu in *Ch'ien-fu lun* 6:4a.

[83] *Lun-heng* 17:14–19a and *passim*.

[84] Etienne Balazs, *Chinese Civilization and Bureaucracy* (Yale University Press, 1964), pp. 194–213.

[85] *Ch'ien-fu lun* 1:5b–6; 3:2–8; 4:5b–9, 15b–17a; 5:10a, 12b. Cf. Etienne Balazs, *Chinese Civilization*, pp. 200–203.

30

INTRODUCTION

between the "inner" and the "outer" worlds, between ideal and reality, and between the superior morality of the Sage and the supreme power of the ruler, as suggested by Wang Ch'ung, ought to be avoided or mitigated. If the Sage, by the strength of his moral character which alone sustains his sagehood, can withstand the pressure of a corrupt society and a wayward ruler, this can preserve the ideal good and prevent a total corruption of the world even at the worst of times.[86] When the conditions are more propitious, the inner light thus sustained by the Sage will shine through to enlighten the outer realm,[87] according to Wang Fu. Although one cannot make the Sage a ruler or the ruler a Sage, one may still hope that the Sage will guide the ruler and the ruler will accept guidance from the Sage. Thus a reunion would be achieved.[88]

This appears to be the assumption shared by those Later Han Confucians who refused appointment to office by the court, or voluntarily resigned or retired form their government posts as an act of protest. Many of them risked their lives or the sufferance of exile in criticizing the powerful relatives of the emperor and the palace eunuchs, who controlled the central government. And they censured those who failed to do so as lacking moral courage and integrity. Their words and actions of protest became the *ch'ing-i*, a purist moral censure against the turbid.[89]

---

[86] *Ch'ien-fu lun* 1:9–10a, 12–14; 2:1b, 3a; 8:13a.

[87] *Ibid.* 2:5b; 3:11b–14a; 7:6–7a.

[88] Although Wang Fu occasionally used the term Sage-ruler (*Sheng-wang*, in *Ch'ien-fu lun* 1:1b, 2:4a, 3:14a), he often separated his discussion of the moral virtue of the Sage and worthy men (*hsien*, also *chün-tzu*) from his discussion of the political accomplishments of the ruler (*ming-chün*, an enlightened ruler, or *jen-chün*, ruler of men); and he considered the two as often in conflict, 1:12–14; 2:1–3, 5b, 6a, 9b, 11b–13; 8:6b–8. An exception was his discussion of the political-moral transformation of the world by the Sage-ruler in 8:10–15, but even here a distinction may be made between the administrative measures of the ruler, which affect the "people's body," and the moral influence of the Sage, which affects "the people's mind and spirit." Wang Fu's ideal was that the ruler should be public-minded (*kung*) in selecting the worthy for office and in taking the Sage as his teacher, 1:1; 2:2a, 3b–9a.

[89] Chi-yun Chen, *Hsün Yüeh*, pp. 18–19 and notes.

INTRODUCTION

The political censure and persecution known as *tang-ku* (the arrest and blacklisting of partisans) ordered by the court against the *ch'ing-i* protesters from A.D. 166 onward only added to the prestige and popularity of the protesters and strengthened their solidarity. The movement received support from ever-widening social circles.[90] The dissidents considered their movement a crusade to assert spiritual and moral leadership over the realm in contradistinction to the political power of the ruling dynasty. They honored their leaders with the titles of "The Three Rulers (*san-chün*), who were worshipped and obeyed by the whole realm," "The Eight Eminents (*pa-tsün*), who were the outstanding leaders of men," "The Eight Guides (*pa-ku*), whose moral conduct was exemplary for the people," "The Eight Aides (*pa-chi*), who assisted the people in following their moral leaders," and "The Eight Treasurers (*pa-ch'u*), who contributed their wealth to relieve the people in plight"—a hierarchy based on moral worth and merit.[91]

While the movement was praised by many contemporary Confucians and later historians as a moral crusade inspired by lofty ideals, motivated by righteous indignation, and sustained by courage and martyrdom,[92] it was in fact complicated by conflicting group interests and violent power struggles inside and outside the government, ranging from the imperial court to the local communities. Its occurrence was, by and large, symptomatic of the antagonism between the centrifugal and centripetal forces at work during the declining phase of the Later Han rule.[93]

The persecution and the crusade came to an abrupt end in A.D. 184, when the Yellow Turbans uprising broke out. Under the threat of the violent Yellow Turbans, the court nullified the measures against the *ch'ing-i* partisans and the

[90] *Ibid.*, pp. 19–30.
[91] *CC* 67 (*lieh-chuan* 57):2–4.
[92] *Ibid.*, 2, 5b, 17a, 19b; 68 (*lieh-chuan* 58):2–3a; 69 (*lieh-chuan* 59):2; 65 (*lieh-chuan* 55):6a. Chao I, *Nien-erh shih cha-chi*, pp. 93–95.
[93] See above, notes 89–90.

INTRODUCTION

latter rallied their support for the court. The insurrection was pacified within a few months, but minor uprisings continued to erupt in various parts of the empire in the ensuing decades. The Later Han dynasty lost much of its power—the former partisans of the *ch'ing-i* movement gained control of the court, while the powerful clans, big landlords, and other consortia of the local élite dominated the provincial areas. Their position was further threatened by the frontier soldiers summoned by the court to fight against the insurgents and by new military leaders emerging from the civil war. The conflict plunged China into a prolonged period of turmoil and destruction. In A.D. 196, the Emperor Hsien and his retinue of high officials were taken into protection by Ts'ao Ts'ao, who revived the Later Han court with a new reign of Chien-an (Establishing Security). The court was at the district town of Hsü, south of Lo-yang. During the next twenty-five years, a titular Later Han dynasty was prolonged while Ts'ao Ts'ao established himself as the indisputable ruler in north China, challenged only by the regional states of Shu (in present Ssuchuan) and Wu (in the middle and lower Yangtse River valley). The Later Han dynasty was formally terminated in A.D. 220 by Ts'ao Ts'ao's son, Ts'ao P'i, who founded the Wei dynasty in a divided China.[94]

Thus when Hsün Yüeh wrote the *Han-chi* and the *Shen-chien* in the early years of the Chien-an era, the mighty Han empire had already come to an end. The emperor whom he served was no more than a precarious figurehead. The country had just experienced a major uprising and was still in the depths of civil war, with a prolonged period of disunion ensuing. During the previous decades, several members of Hsün Yüeh's extended family in Ying-ch'uan (about one hundred miles southeast of the Later Han capital Lo-yang, in present Honan province) had actively participated in the *ch'ing-i* crusade and suffered a flurry of court persecutions. Hsün Yüeh's uncle, Hsün Shuang (A.D. 128–190) "had been persecuted

---

[94] Chi-yun Chen, *Hsün Yüeh*, pp. 30–47.

INTRODUCTION

under the *tang-ku*; he lived in hiding in a coastal area and later fled south to the Han riverbank; for more than ten years he devoted himself to the task of writing and became acclaimed as a versatile Confucian." Hsün Yüeh's cousin Hsün Yü, meanwhile, had actively cooperated with Yuan Shao, Ts'ao Ts'ao, and a former student leader of the Imperial Academy, Ho Yung (d. A.D. 190) in a secret mission to provide assistance to the persecuted partisans of the *ch'ing-i* crusade. While Hsün Yüeh himself did not participate in the *ch'ing-i* partisan activities, he was nonetheless affected by a court-ordered persecution from A.D. 176 onward when the central government extended the persecution to all relatives of the partisans "within the five mourning grades." He spent his early adult life from about the age of twenty-nine in provincial obscurity.[95]

The crusade and persecution propelled many leaders of the *ch'ing-i* into national prominence. When the partisans gained control of the court after the Yellow Turbans insurrection, Hsün Shuang was promoted to the high office of *Ssu-k'ung*, a ducal minister, and Hsün Yü became Prefect-Protector of the Palace (*shou kung ling*). The home town of the Hsün clan in Ying-ch'uan was soon ravaged by rampaging soldiers, and Hsün Yü led his clansmen to join the allied armies of Yuan Shao and Ts'ao Ts'ao. As civil war continued, Hsün Yü became Ts'ao Ts'ao's principal adviser and was the architect of the restoration of the Later Han court of the Chien-an era. It was probably through his recommendation that Hsün Yüeh was appointed attendant to the figurehead Emperor Hsien and the Custodian of Secret Archives at the titular court.[96] This experience was important to Hsün Yüeh as an observer of historical and contemporary issues, and his position made him a quasi-official spokesman for the Confucian literati at the Chien-an court.[97]

o o o

[95] *Ibid.*, pp. 24–29, 66–70.
[96] *Ibid.*, pp. 41, 44, 49, 60, 75–83.
[97] See note 1.

INTRODUCTION

At the beginning of this introduction, we traced the tension between moral dogmatism and philosophical skepticism in Hsün Yüeh's writings as symptomatic of a crucial transition between the Han and post-Han periods. From a broader perspective, we examined the complex of agnostic-skeptic and positivistic-dogmatic attitudes and the polarity of concerns over the "inner" and the "outer" realms in pre-Han and Han thought. The evolving intellectual legacy paralleled the changing socio-political reality from the disorder of late Chou times through the rise of the Ch'in-Han imperium to its breakdown in Hsün Yüeh's time. This evolution of thought involved not only the problems of ideology, morality, and cosmology, but also some fundamental questions concerning the role of human reason, the basis of knowledge, and the meaning of life.

The rising optimism of the late classical and early Han thinkers with regard to the efficacy of human reason and action led them to ignore many of the fundamental questions that had been raised but left unanswered—or inadequately answered—by the early classical thinkers. It was not until later Han times, when the constrictive grip of the imperium and the prospect of its total collapse aroused a keen awareness of the human predicament and a sense of profound disillusionment, that thinkers like Wang Ch'ung and Wang Fu began to critically reexamine the legacy of classical and Han thought. Ironically, both Wang Ch'ung and Wang Fu still took the certainty of superior human rationality for granted in their critique of Confucianism and of the state and society of their own times. It was Hsün Yüeh who, in a rather abrupt and obtuse way, laid out in his writings many of the unanswerable and therefore utterly disturbing questions.[98] Herein lies the

---

[98] Although both Wang Ch'ung and Wang Fu were pessimistic about the outside world, they did not lose faith in the "inner realm" as realized by the Sage. They criticized the "vulgar Confucian teachings," but did not question man's ability to attain true knowledge. In comparison, Hsün Yüeh seldom explicitly criticized the Confucian teachings, and many of his remarks appear to be quite innocuous or apologetic at first sight. However, as the following

intrinsic merit of his ideas as well as his contribution to China's intellectual legacy.

As discussed earlier, the moral and political dogmas of Han Confucianism were based primarily on the interpretation of history and cosmology. These were examined by Hsün Yüeh in two of his longest discourses (*lun* 論) in the *Han-chi*. In his first discourse in the *Han-chi* (Selection One), Hsün Yüeh discussed in detail the complex factors determining the outcome of historical events. He classified these factors into three broad categories: general conditions, specific situations, and the state of mind. He cited many historical events to demonstrate how these factors might differ in some apparently similar historical circumstances. His argument, pushed to its logical conclusion, would make every event unique in itself, thus frustrating any attempt to draw a parallel or "lesson" from history. He ended the discourse by stating the futility of using history to predict the future.

In another long discourse in the *Han-chi* (Selection Two), Hsün Yüeh analyzed the problem concerning "the correspondence between the cosmos and men." At the outset of the discourse, he seemed to reaffirm the Han Confucian belief in the cosmological theory of correspondence. He mentioned that men's actions had far-reaching effects and would produce an ultimate echo from Heaven; men were capable of knowing this and of controlling the result by changing their conduct. These views, together with Hsün Yüeh's metaphor of the object (man and his actions) and its shadow (natural or cosmic apparition), or a sound and its echo, tended to support the anthropocentric world-view suggested by Confucius and advanced by Hsün Tzu, which was the basis of Han Confucian cosmology. Thus it seemed that, while Wang Ch'ung had used historical study to repudiate cosmology, Hsün Yüeh questioned the validity of historical knowledge but reaffirmed the possibility of cosmological knowledge.

---

discussion will show, the questions he raised are so radical in a deeply philosophical sense as to make him un-Confucian.

INTRODUCTION

But this is illusionary, for the main thrust of this discourse was the incomplete and imperfect state of man's knowledge, which Hsün Yüeh saw as a problem in the views of both the proponents and opponents of cosmology. He intimated that the question of cosmological knowledge is far more complicated than that of historical knowledge. If Hsün Yüeh was less sanguine than most Han Confucians about the certainty of historical knowledge, he would be the least sanguine about the attainment of cosmological knowledge. He cautioned that one must be flexible and versatile in learning, for even with maximal human versatility what one could achieve was not exactitude but only an approximation of truth. Hsün Yüeh classified the relationship between nature and man into three general categories: that which is realized solely by nature with no need for human effort; that which must be accomplished by man and cannot be accomplished without human effort; and that which cannot be accomplished in spite of the greatest human effort. The first and last categories were the province of fate (including man's innate limitations as discussed by Wang Ch'ung). The middle category represented the uncertainty and the promise of human endeavor. Thus Hsün Yüeh counselled that the "superior man" should exert his mental power to the utmost and then resign the rest to fate. It is in this sense that Hsün Yüeh affirmed both the importance and the limitation of human knowledge and action.

In the same discourse, Hsün Yüeh presented his ideas on the limitations of man's knowledge in yet another way. He stated that "the way of Heaven and the Way of Men have their similarities and their differences." He advised that one should study these differences first and then search for the similarities, and not the other way around. He seemed to suggest that knowledge based on the concrete and the particular (different objects) is more reliable than that at the generalized, abstract, and inferential level (i.e. inferred similarity). Hsün Yüeh supported this contention by the three kinds of examples that he gave in analyzing the complicated interaction between nature and man. He began with the concrete example of the human body or, more specifically, bodily disease, then moved on to the less material effects of education, and finally discussed the abstruse Way of Heaven.

Hsün Yüeh adopted a similar procedure in the scheme of his *Shen-chien*. By entitling his second work *Shen-chien*, he seemed to mean it to be an expansion (*shen*) of his reflections on history (*chien*). Although the word *shen* also had the meaning of "reiterating," and some of the passages in the *Shen-chien* tended to repeat Hsün Yüeh's *lun* discourses in the *Han-chi*, the *Shen-chien* was not primarily a reiteration of his historical reflections in the *Han-chi* or elsewhere. In writing the *Shen-chien*, Hsün Yüeh was drawn further and further into abstruse philosophical issues. But even this seems to accord with his scheme of reasoning from the concrete to the abstract: in the *Shen-chien* the chapter on government (*SC* 1) is followed by chapters on current affairs (*SC* 2), common beliefs (*SC* 3), and finally philosophical dialogues (*SC* 4–5).

Hsün Yüeh's commitment to tackle problems that were abstract and abstruse may be found in his discussion of the three meanings of *chien*: a mirror reflecting one's appearance, people reflecting human virtue, and history reflecting the bygone ages. He specifically warned that "those who only look at the mirror will fail to see true reflections" (*SC* 4.2). Observing that "a person wearing fine clothes would not wallow in the dirt," he lamented, "how superficial is he who cares for his clothes but not his manners, or he who cares for his manners but not an enlightened spirit" (*SC* 5.1). Thus, in spite of his view that human knowledge is more reliable concerning the concrete and the particular, Hsün Yüeh did not lose sight of the more elusive meaning of reality, nor flinch from his search for it.

A solution to this seeming inconsistency may be found in his discourse concerning "the correspondence between the cosmos and men" in the *Han-chi*. There Hsün Yüeh intimated that man's observation and experience were often fixed at a specific time, while in reality everything is in a continuous process of change. Thus, although that which has not yet come into being cannot be seen or known, it would be folly to deny its meaning on the basis of one's observation and experience of that which has been. This raises a crucial issue concerning the relationship between idea and reality in classic Chinese thought. On this issue Hsün Yüeh cautiously reaffirmed the nominalistic attitude of the Sophists (*Ming-chia*). He not only asserted that one should know the concrete and the particular and then infer the general and abstract, as discussed earlier, but he also indicated that speech, words, and names are but the medium that communicates what is real and true; if reality and truth were complex and abstruse, the words of the Sage could not but become complicated and difficult (*SC* 5.12). According to this view, the so-called "rectification of names" should consist of using names appropriate to reality and changing a name when it is found to be inappropriate, not forcing reality into the straitjacket of names and ideas, as many Confucians understood the concept (*SC* 5.18).

Even though names and ideas are thus grounded in the concrete and the particular, they themselves are abstract and general, and as such often fail to do justice to the complicated and dialectical structure of reality. Hsün Yüeh presented several examples of the ambiguous nature of words: there are "crimes that constitute real crime" and "crimes that constitute no real crime"; there are "troubles that beget more troubles" and "troubles that mitigate troubles" (*SC* 4.7); what is considered "indistinct" may in fact be "distinct" (*SC* 4.6); what is known as "yielding" is in fact "winning" (*SC* 4.9); and what is called "fortunate" may turn out to be "unfortunate" (*SC* 4.12).

To Hsün Yüeh, doctrinaire inflexibility, simple-mindedness, and one-sided discrimination are most detrimental to enlightened human understanding. Conversely, flexibility, versatility, and a tolerant and all-embracing attitude are sagely virtues. He saw the need to be open to many paths toward the truth— one must lay a net consisting of many meshes so that one of them may trap a bird (*SC* 2.19). In other analogies, he pointed out that it takes different notes and tunes to make good music and different flavors to produce good food: "To prepare food by flavoring water with water, who can bear to eat it?" (*SC* 4.19.)

INTRODUCTION

Even within the Confucian tradition itself, he remarked, there were numerous controversies and questions that could not be settled; "the different opinions could not all be correct," but "if one compares and discusses them, there may be something worth learning" (*SC* 2.18).

At times Hsün Yüeh seemed tacitly to accept Chuang Tzu's view of extreme relativism. Overall, however, he came closer to the original position of Confucius, who—unlike Chuang Tzu—recognized that, as human beings living in society, we cannot dispense with discriminative knowledge and moral judgment. Subjective as they may be, they are part of human nature. Thus flexibility and tolerance are for Hsün Yüeh not the same as indiscriminate acceptance, unprincipled versatility, or abrogation of judgment. For example, he criticized the misuse of cosmology in interpreting omens and disasters (*SC* 3.2–3)—although he did not deny the far-reaching moral, natural, and cosmic effects of human actions (*SC* 4.18). He also denounced many superstitious practices, but at the same time tried to offer some rational basis or "religious meaning" for them (*SC* 3.1, 3, 4–5, 7–9, 12–13). He questioned the authenticity of the apocryphal-prognostic *wei* books that had been favored by the Later Han court and posed a serious threat to rational Confucian thought, but when someone asked whether these books should be burned, he answered: "Although we cannot say that these books are Confucius' writing, there are still things in them worth learning. Why should we burn them?" (*SC* 3.15). The same broad-minded attitude underlay Hsün Yüeh's counsel on the need to preserve a detailed historical record of the various actions and words of as many men as possible—all the way from the ruler down to the common people (*SC* 2.22).

A central contradiction in Hsün Yüeh's writing is that despite his disdain of dogmatism he appears to be highly doctrinaire in upholding the Han Confucian political and moral ideas concerning the ruler and the court. Here Hsün Yüeh's ideas differed from those of his predecessors only in nuances.

INTRODUCTION

According to Hsün Yüeh, the empire is a public institution, hierarchic and yet holistic. While the emperor reigns supreme in the imperium, he exists to serve the multitude of people who constitute the foundation of the state (SC 1.4, 32, 33, 4.4). The emperor is not one of the people and it does not behoove him to conduct himself like an ordinary person. An ordinary person is a private individual; he has his feelings of likes and dislikes; he may be wise or stupid, good or bad; he may be selfish in profit-seeking or he may be honest in following the path of righteousness; and he should be reformed by education and controlled by law (SC 1.9–12, 34–38; 2.3, 9, 11–13; 4.13; 5.14–21). But it is not so for the emperor. The emperor is not a private individual. He has the most awesome power and responsibility. The possession of the realm is not a pleasure but a heavy burden (SC 1.39). For in eight out of nine conceivable situations the state may be endangered or in decline (SC 1.17), and orderliness is a precarious equilibrium which can be sustained only with great effort (SC 1.16–17). The ruler must make a great effort to restrain himself so as to expand the happiness of his people (SC 1.39). It does not suffice that he loves his people as his own son, nor even that he loves them as much as he loves himself; he ought to love them more than he loves himself—for the state survives only as long as the people survive (SC 4.4.).

To do this, the emperor must be entirely selfless. He owns the empire but is himself owned by it. He ought to have no private expenditure nor personal resources; his every need is provided for by the public as part of governmental expenditures supported by public taxes, levies, and corvée services; he has power but not privilege; he may grant official rewards but not personal favors (SC 1.34). Since his emotions, thoughts, and actions have far-reaching effects on the whole realm, they should all belong to the public. In this extreme sense, the emperor is not entitled to his private feelings of like and dislike. For what he likes his people will also like and what he dislikes his people will also dislike—this is influence by example, the essence of moral education. Furthermore, what the emperor likes he encourages through reward and what he dislikes he discourages by punishment—this has the force of law. Therefore, the emperor ought to set aside his personal likes and dislikes and follow only the criteria of what is good for the public (*SC* 1.3, 10, 13; 4.13).

In order to accomplish this goal, the emperor must discipline his mind and regulate his conduct according to the highest standard of self-sacrifice set by the former Sage-kings (*SC* 4.10–11). His daily life should be patterned by the strict regimen of court etiquette; his every movement should be subject to ministerial guidance, open to public scrutiny, and entered into the historical records (*SC* 1.29, 2.22, 4.3). And he must exercise strict control over those surrounding him, i.e., members of his family and attendants in the palace, so as to prevent them from currying his personal favor, corrupting his public spirit, or misusing his authority (*SC* 1.30; 2.17, 21; 4.16, 17). The same principle applies to the court ministers and officials, except that while the relatives and attendants of the emperor are considered his personal followers the court ministers and officials serve the emperor mainly in their public capacity and are not his personal following (*SC* 1.4, 16, 18–26, 41; 2.11; 4.3, 8, 15–16).

The court ministers and officials are rather like the colleagues of the emperor in their cooperative effort to follow the Way, which is the guiding principle and the moral foundation of the state (*SC* 1.1, 4). The emperor should understand and pursue the Way (*SC* 1.10, 31; 4.7), and lead his people to it (*SC* 1.33, 39–40). The court ministers and officials serve the ruler only on account of the Way; they must be prepared to disobey a ruler who deviates from the Way (*SC* 4.7, 15), and the ruler ought to humble himself and yield to his ministers and officials on account of the Way (*SC* 1.39; 4.9, 39).

Thus, although Hsün Yüeh upheld the Confucian ideal of the Sage-ruler, his emphasis was one of negative restraint. The image of an ideal ruler, as projected in the *Shen-chien* and some of the discourses in the *Han-chi*, is one held in captivity by the imperium rather than the true master of it. The different nuances of Hsün Yüeh's demands on the ideal ruler and on the

INTRODUCTION

ideal official are: the ruler should first concern himself with the Way, then with his benefit to the people, and lastly with his personal virtue; on the other hand, an official should concern himself first with his personal conscience, then with his official duty, and ultimately with the Way (*SC* 1.40–41). What is implied here is that an official, who is born a private individual and acquires his position according to the Way, is answerable to his personal conscience and entitled to his own righteous feelings, whereas an emperor who is born into his position is not.

Hsün Yüeh's attitude toward the ruler may be partly reflective of the early Han Confucian reservation about the imperial system, partly the result of his experience in the *ch'ing-i* movement and the *tang-ku* persecution, and partly based on his observation of the actual condition of the Chien-an reign. As an historian and high official at the titular Han court, Hsün Yüeh was well aware that his lofty ideals and high-sounding principles concerning the ruler and the state could not be realized in his own time nor in the near future.[99] In contradistinction to such lofty ideals and principles was Hsün Yüeh's counsel for flexibility in government policies and programs, in which he advocated that one must act according to what is possible, timely, and expedient in the complicated and ever-changing situations that confront a person (*SC* 2.2, 7, 9, 11, 20).

It is unlikely that Hsün Yüeh failed to notice the contradiction between his adherence to the ideal and his counsel for expedience, or the tension between the dogmatic and the skeptical strains in his thought. It seems more likely that he considered the ambivalence as essential to a tolerant and all-embracing intellectual attitude. After all, would it not be hypocritical for one to tolerate the apocryphal-prognostic *chan* and *wei* books or Taoist superstitions but not the Confucian ideals? Dogmatic as the latter might be, it could be that "one mesh of the net" that catches the high-flying bird. If one were

---

[99] Like many historians before him, Hsün Yüeh addressed his work primarily to future generations of readers; see my translation of Hsün Yüeh's memorial with regard to the *Han-chi*, *Hsün Yüeh*, pp. 90–91.

INTRODUCTION

to preserve an historical record of as many of the actions and words of men as possible, would it not be imperative to preserve and reiterate the ideals of the Sage-kings, lest they be totally lost to future generations?[100]

The attitude of considered ambivalence also characterized Hsün Yüeh's discussions of the moral situation, the "inner" and "outer" realms of reality, and the issues concerning fate (*ming*), human nature (*hsing*), and feelings (*ch'ing*), intention and determination (*hsin* and *chih* 心, 志), and moral action (*hsing* 行). At the outset, it should be noted that although Hsün Yüeh often mentioned the Confucian concept of "three grades of men"—the highest being the Sages, the middle being ordinary human beings, and the lowest being the incurable morons—his interest was mainly directed towards the morality of ordinary men. The Sage is born with knowledge and goodness and can fulfill his moral potential without effort, and the moron is incapable of any of this. Therefore, the discussion of complicated moral issues and the role of moral education is superfluous for them. Moreover, since the Sage, as idealized by Wang Ch'ung and Wang Fu, is devoted to the inner world and unaffected by external forces, and by inference the moron is completely controlled by outside influences, only to an ordinary person would analysis of the interaction of the inner and outer worlds in relation to the resolution of specific moral situations be relevant. Because "the highest and the lowest of men ... cannot be changed," what Hsün Yüeh was interested in was the malleable mass of men (*SC* 1.11, 35; 2.3, 9; 4.1; 5.15, 19; also *Han-chi*, Selection Two).

Hsün Yüeh's view of the inner and outer dimensions of human existence and morality was highly complex. In certain respects he asserted the primacy of the internal over the external aspects of being. He stated that "inner defense" is more crucial than "outer defense" for the well-being of the human body as well as of the state (*SC* 4.17), just as "inner feeling" is more important

---

[100] *Ibid.*, pp. 85–86, 90–91. See translation of *SC* 1.1; 2.18–19, 22 below. Hsün Yüeh wrote: "Highest is the one who does not discriminate the past and the present" (*SC* 5.23), which certainly includes the ideal of the former Sage-rulers and the precepts of Confucius.

than "outer decorum" as a quality of the "superior man" (*SC* 3.3; 5.3). Even though the "cosmic environment" may be likened to one's abode, Hsün Yüeh maintained that it is the inner worth of a man that determines how well he relates to the cosmos (*SC* 3.3). However, to Hsün Yüeh, man's inner world includes his entire being as a "centered" personality. It is shaped by his moral choices and conduct, which are the products of the web of interaction of internal and external factors.

Some of these factors are predestined; others are open to the formative power of the individual. Hsün Yüeh thought that fate (*ming*) plays an important role in human life and is beyond human control—a position similar to that of Wang Ch'ung. On the other hand, unlike Wang Ch'ung, he believed that human effort could make a great difference in what happens to a person (*SC* 5.14, 15). Therefore, instead of abjectly describing the overwhelming power of fate, Hsün Yüeh delved into the subtleties of its encounters with man's individual will and abilities.

Hsün Yüeh admitted that fate determines the quality of man's innate nature (*hsing*) (*SC* 5.14). But in line with his interest in that which is observable and changeable, he did not dwell on the question of whether innate human nature is good or bad—a question which had occupied the minds of many Confucian thinkers. Hsün Yüeh devoted only five sentences to reviewing the opinions of his predecessors on this issue, and then spent another paragraph criticizing their arguments, which he considered "not very well reasoned" (*SC* 5.15). Instead of concentrating on innate human nature, he paid more attention to the role of feelings (*ch'ing*) in determining man's action. He adopted the critique of Liu Hsiang (77–6 B.C.) aimed at Tung Chung-shu's theory that human nature is good and human feelings are bad. Hsün Yüeh was appalled by the thought that if one asserted that human feelings are categorically bad, the Sage would have to be characterized as devoid of feelings (*SC* 5.15).

Whereas innate human nature is man's essence or potential in an inanimate state, and hence it is non-observable, Hsün Yüeh viewed human feelings as being the initial expressions of

INTRODUCTION

human nature in the form of observable likes and dislikes. Since there are as many bad persons as there are good ones, one can infer from this that man's essence or potential is both good and bad. However, it can also be argued that innate human nature is either purely good or purely bad, since it cannot be seen. In contrast, feelings of like and dislike—being observable—can be correctly linked to the observed good and bad actions of men. Therefore Hsün Yüeh concluded that the judgment of good and bad can only be made with respect to accomplished actions, and that in such cases an objective moral standard is applicable (*SC* 5.16).

Hsün Yüeh's criterion for moral judgment is thus based on external manifestation of evil. He argued that a person should not be considered evil merely because he has an evil nature (i.e., a potential and inclination) that is dormant, or because an evil thought has arisen in his mind which has not yet generated an evil action, or even after he has set an evil action in motion but has reversed it before it has gone too far. Such shortcomings, Hsün Yüeh remarked, are common in the ordinary nature of men. Only when one completes an evil action should one be condemned as evil (*SC* 5.21). For this reason, "the superior man honors propriety of conduct but does not question motives," especially in a corrupt age (*SC* 5.22)—a rather un-Confucian view.

Hsün Yüeh's emphasis on the external with regard to moral questions appears to have reached an extreme in his comparison of the good (benevolence and righteousness, *jen-i*) and the bad (selfish profit-seeking) with "meat and wine." These are competing attractions, each appealing to man's feelings; "that which wins will have its way" (*SC* 5.17). According to Hsün Yüeh, this fact disproves the theory that "by feelings (or desire) man seeks profit and by nature man seeks righteousness." The analogy seems to suggest that he considered man's morality to be determined mainly by external forces—the meat and wine of the world, to which man passively responds. If this interpretation is correct, Hsün Yüeh's view of man's moral situation was far more pessimistic than Wang Ch'ung's.

INTRODUCTION

It may be argued that Hsün Yüeh's metaphor of meat and wine implies that the competition is between man's inner goodness (innate moral nature, *hsing*) and outside temptation. This would be in accord with the view of the majority of Confucian thinkers, especially the neo-Confucians of Sung, Ming, and Ch'ing times.[101] But Hsün Yüeh emphatically stated that this was not what he meant. According to him, goodness (*i*, righteousness) and badness (*li*, selfish profit-seeking) ought to be compared on the same plane, as both external or else both internal (Hsün Yüeh did not explicitly discuss the second possibility because what is internal cannot be observed), but not one internal and one external, for in so doing one has already prejudged the issue (*SC* 5.16, 18). In this sense, Hsün Yüeh greatly emphasized the external realm.

However, Hsün Yüeh asserted that good and bad, or meat and wine, have to compete with one another only when a person's feelings draw him to both. He contended that "if a person likes only one [and not the other], then even though he is able to take both, [he will not]" (*SC* 5.17). Thus, in orienting one's feelings of like and dislike, a person can exercise his will power and make a choice (*SC* 4.5; 5.14, 18, 21). Since innate human nature (*hsing*), like fate (*ming*), is determined by nature or Heaven, over which man has no control, what man can control is his own feelings (*ch'ing*) of like and dislike. This becomes a crucial factor in Hsün Yüeh's concept of self-mastery, and hence of man's moral autonomy (*SC* 4.11; 5.2-3).

[101] According to Mencius, the conflict occurs mainly between man's innate goodness and a harmful environment. Although Hsün Tzu contended that human nature is evil and that goodness is the result of environmental constraint (i.e., cultural influence and social pressure), he also stated that man desires goodness because of his badness. This implies an inner conflict between man's evil nature and his moral need. It was in neo-Confucianism that the conflict between "Heavenly reason" (*t'ien-li*) or "innate human nature" (*t'ien-hsing*), on the one hand, and "human desire" (*jen-yü*) or "emotion and desire" (*ch'ing-yü*), on the other, was considered a basic element of the moral situation. Fung Yu-lan, *History*, Vol. 1, pp. 122, 124, 127, 280–288; Vol. 2, pp. 516, 518, 553, 555, 556–561 and *passim*.; *A Short History of Chinese Philosophy* (New York, 1960), pp. 70, 144–147, 283, 287–289, 301–303. Thomas A. Mertzger, *Escape*, p. 108 and *passim*.

INTRODUCTION

This concept is not entirely new. Both Wang Ch'ung and Wang Fu had noted the important distinction between what is predetermined by Heaven (i.e. *hsing* and *ming*) and what can be decided by man (human effort and action). However, instead of elaborating on this critical issue, Wang Ch'ung and Wang Fu devoted their attention mainly to the discussion of the predetermined or heteronomously determined factors and how these affect man's life. Although they postulated the importance of self-determination in man's moral autonomy, they did not explicitly discuss how this can be achieved. Somehow, in their writings, they confused the issue of man's moral autonomy with the mystery of sagehood and a subjective "inner realm," thus reifying the inner and the outer dimensions of reality into two dichotomous regimes. Hsün Yüeh also emphasized the difference between what is observable (the external) and what is non-observable (the internal in man), but he considered the two as a continuum of the inanimated and the animated stages of the human mind; since the two cannot be clearly demarcated, the controversy over the priority of one or the other appeared to him as beclouding the moral issue (*SC* 5.15–16, 18).

Hsün Yüeh's discussion on human feelings, which appears to be disproportionately lengthy in the present *Shen-chien*, was singled out by the Sung scholar Huang Chen (fl. 1256–1270) for severe criticism.[102] It should be noted that elements of the emotional were important in Confucius' view of life and human-heartedness (*jen*) and in Mencius' concept of man's innate nature.[103] In this respect, Hsün Yüeh's emphasis on human feelings was more congenial to the pristine Confucian attitude

---

[102] See below pp. 59, 63–64.

[103] The *Analects* begins with Confucius' statement on "joy" and "anger," Legge, I, p. 137. The cardinal Confucian virtue, *jen*, is commonly rendered into English as "love" or "kind-heartedness." Mencius considered man's instantaneous feelings of commiseration, shame and dislike, humility and indignation, to be "proof" of man's innate goodness, and man's compassionate heart as the basis for the politics of compassion. Fung Yu-lan, *History*, Vol. 1, pp. 120–121; Legge, II, pp. 138–144.

than was the neo-Confucian emphasis on the contrast of good and evil in the dichotomy between human nature and human feelings.[104] Although the views of Confucius and Mencius on human virtue were strongly oriented to the emotive, affective, and aesthetic, their discussions of human feeling were implicit and accidental. It was Hsün Yüeh who explicitly discussed the important role of human feelings in character formation and personality integration. His attitude had considerable impact on the post-Han intellectual and literary movements.[105] His insight is even more meaningful in the light of modern psychoanalytic findings.

Hsün Yüeh paid special attention to the issue of human feelings because, first, they are observable; second, they represent an interaction between the internal and the external realms in human life; and third, they are crucial for man's attainment of self-mastery. According to Hsün Yüeh, to achieve self-mastery one needs a firm and persevering will (*SC* 4.10–11; 5.2, 14)—unbridled by hesitation, fear, or confusion (*SC* 4.5, 5.3, 13). This explains why Hsün Yüeh strongly opposed Tung Chung-shu's theory that in order to be good a person must use his moral nature (*hsing*, conscience?) to suppress his feeling or desire (*SC* 5.17)—for this might create tremendous inner conflict and consequently weaken one's strength of character. For the same reason, he considered the feeling of shame (*ch'ih*) to be important but not fundamental to one's moral goodness (*SC* 5.24). He asserted, "One's ideal should come naturally from oneself; what does it have to do with shame?" (*SC* 5.25).

This last statement, together with his metaphor of "meat and wine" discussed earlier, does point to a serious problem in Hsün Yüeh's moral philosophy, i.e., his reticence on the issue of the higher and lower forms of human emotion, *ch'ing* (human feelings) and *yü* (instinctual desires). The distinction between the refined, aesthetic, and spiritual sentiments, on the one hand, and crude, violent emotions, on the other, became crucial in the

---
[104] See note 101.
[105] See Fung Yu-lan's discussion of neo-Taoist sentimentalism in *Short History*, pp. 231–240.

INTRODUCTION

post-Han *ch'ing-t'an* movement. And the difference between the "morally pure and articulated feelings" and the "untrammeled flow of desires" was basic to the neo-Confucian conception of human morality.[106] But even here Hsün Yüeh's reticence may be considered intentional. As in the *Republic* of Plato as well as in the post-Han *ch'ing-t'an*, the discrimination between the finer and the cruder forms of human affections suggests a strongly aristocratic orientation. On the other hand, the term "likes and dislikes" used by Hsün Yüeh was non-hierarchical and consistent with his attitude of all-inclusiveness discussed earlier. Sometimes, this term, together with the metaphor of "meat and wine," conjures an image of the ordinary "man in the street," so to speak. This reinforces our previous argument that what interested Hsün Yüeh most was not the highest Sage nor the lowest moron but the majority of average men. In this connection, it may be remarked that, whereas the concern of Wang Ch'ung and Wang Fu over the inner vis-à-vis the outer realms had produced a model of the Sage versus the ruler, Hsün Yüeh's emphasis on the strength of character and personality integration called for a paradigm suitable to all ordinary men and susceptible to the influences of both the Sage and the ruler.

For such ordinary men, education and legal restraint are necessary (*SC* 1.3, 11, 35–36; 2.9; 5.19–20). Although Hsün Yüeh hesitated to pass judgment on the unobservable inner world of men (hence the skeptical strain in his thought), as a Confucian historian his "praise and blame" of men's good or bad, successful or unsuccessful, and fortunate or unfortunate actions—the observable deeds and historical events—were explicit and clear in both the *Han-chi* and the *Shen-chien* (hence the dogmatic strain in his thought). As Hsün Yüeh intimated in his first discourse in the *Han-chi*, such historical events will not repeat themselves. But, once they come to pass and their consequences become known, they can be judged. These judgments constitute the lessons of history. They guide the ruler in the administration of rewards and punishments and inform the

---

[106] See notes 101 and 105.

ordinary man as to how to conduct himself. Although Hsün Yüeh asserted that one's ideal should come naturally from oneself, it was mainly from the good historical examples that the ideal acquired its content and became firmly set (*SC* 2.22; 4.2–4, 6, 9–12, 15; 5.25). In appealing to external authority—the regiment of the ruler and the example of the Sage—to adjudicate and inform human morality, Hsün Yüeh remained in the orthodox Han Confucian fold.

Unlike post-Han thinkers of the *ch'ing-t'an* movement, Hsün Yüeh did not abandon the external world. He did not endorse the pursuit of an inward search for purely subjective meanings. He declared, "To know oneself, the search is inward and is easier; to know others, the search is outward and is more difficult; knowing oneself is not the same as knowing others." This is one of the most un-Confucian statements made by Hsün Yüeh. It contradicts the common Confucian thesis that self-knowledge and self-mastery are difficult, and that with self-knowledge and self-mastery one acquires a euphoria and a mystique (*te* 德) by which one can know or even control others.[107] The statement is consistent with Hsün Yüeh's attitude of skepticism, which in this case is directed toward the external world and tended to counterbalance his many apparently dogmatic pronouncements therewith.

According to Hsün Yüeh, the attainment of knowledge and moral virtue requires a combination of inner and outer strength. However, even with the utmost effort, he intimated, ideals remain ideal and differ from reality. What can actually be achieved is no more than an approximation of the ideal, just as what can actually be known is but an approximation of the truth and the best political regime that can actually be founded is no more than a state of ordinary peace. "But if only this can be approximated, there will be no evil—this is good enough" (*SC* 5.25). With this pessimistic note, Hsün Yüeh concluded his *Shen-chien*.

o o o

[107] See Thomas A. Metzger, *Escape*, p. 264, note 243.

INTRODUCTION

For some eight hundred years, from the Age of Disunity to the early years of the Sung dynasty, Hsün Yüeh was hailed as one of the most outstanding Confucian historian-thinkers.[108] As the foregoing analysis shows, his works are important not only for their historical value in reflecting the crucial transition from the Han era to the post-Han era, from the age of Confucian dominance to the age of Taoist and Buddhist ascendance in Chinese history and thought, but also for their intrinsic intellectual merits. Even from the fragmented version of the present *Shen-chien*, we can see that Hsün Yüeh raised many of the fundamental questions in Chinese thought. Some of these questions had been left unanswered by the pre-Han thinkers and were glossed over by many Han Confucians. There were also questions which arose from the reality of the new empire that were unanticipated by the pre-Han thinkers but became critical in late Han and post-Han times. Hsün Yüeh tried to unravel these difficult issues as best he could. If sometimes he appears to lack originality, it is because he dared to reopen those questions which had been asked time and again by his predecessors without satisfactory answers. Some of these questions allow no definite answers, and Hsün Yüeh had the temerity to venture his discussion of them in that light. His originality lies in the manner in which he tackled these issues, rather than in their solution. Given the terseness of his style and the fragmentary condition of the present *Shen-chien*, one is impressed by the scope and the intricacy of the problems that he dealt with in the book.

By and large, Hsün Yüeh may be considered one of the most un-Confucian of Confucian thinkers in traditional China. He did not explicitly denounce or renounce Confucianism; in fact he often defended Confucianism in a highly dogmatic and rhetorical way. But the total configuration of the issues he raised was ominous to the Confucian orthodoxy in old or new forms. For eight long centuries after his death, Hsün Yüeh remained in good Confucian graces. This was

[108] Chi-yun Chen, *Hsün Yüeh*, pp. 171–173.

partly due to the high prestige of the Hsün family in the aristocratic establishment during the Age of Disunity and T'ang times[109] and partly because philosophical Confucianism was at low ebb and a dominant Confucian orthodoxy was non-existent during this period.

It is no accident that Hsün Yüeh's works fell into disfavor with the rise of neo-Confucianism, especially the philosophical-speculative strain of neo-Confucianism, after mid-Sung times. Both in form and in substance, Hsün Yüeh's thought was the opposite of neo-Confucianism. Agnosticism and skepticism were anathema to the neo-Confucian attempt to define an ultimate Confucian doctrine or to establish a new orthodoxy. Historical learning and cosmology, whose validity Hsün Yüeh questioned, were considered by many neo-Confucians to be fundamental to their new orthodoxy. Hsün Yüeh's attitude of considered ambivalence—with regard to the dynastic ruler, to morality, and to the relationship between the "inner" and "outer" realms—was a threat to the spirit of absolutism of the late Sung, the Ming, and the Ch'ing dynasties. Thus for some eight hundred years (from the eleventh century to the present), while Hsün Yüeh's name remained ingrained in history, his ideas were conveniently relegated to oblivion.[110]

---

[109] *Ibid.*, pp. 162–175.

[110] The *Shen-chien* was routinely listed in the bibliographical works of this period as a Confucian (*Ju-chia*) "philosophical writing" (*tzu*). Its only commentary was written by a second-rate scholar in the sixteenth century. See the discussion in the next chapter. The *Han-chi* was considered by many traditional scholars as merely an abridgement of the *Han-shu* and valuable only for textual collation with the *Han-shu*. All these turn out to be only half-truth. See Chi-yun Chen, *Hsün Yüeh*, pp. 2–3, 175–177. For Hsün Yüeh's historiographical accomplishments, see *ibid.*, pp. 84–126.

II. Textual Problems of Hsün Yüeh's Works: The *Han-chi* and the *Shen-chien*

Important as they are, neither the *Han-chi* nor the *Shen-chien* has been the subject of critical detailed study by traditional or modern scholars.[1] With regard to the *Shen-chien*, the work had once been considered of such significance that it received special mention in both Yuan Hung's (328–376) *Hou Han-chi* (Chronicles of the Later Han) and Ssu-ma Kuang's (1010–1086) *Tzu-chih t'ung-chien* (Comprehensive Mirror for Aid in Government)—histories which seldom included the titles of scholarly or literary works.[2] But the *Shen-chien* was on the whole neglected by Chinese scholars after the middle of the Sung dynasty; some even suspected it to be a forgery. When the first annotated edition of the work was produced and printed in the Ming dynasty (see edition no. 2 of the *Shen-chien* in the following discussion), it was considered by those who contributed a preface to the edition to be a subversive teaching.[3] There are many uncertainties regarding the history of the transmission of the text of the *Shen-chien*, and the present version is extremely corrupt.

As the following study shows, the *Shen-chien* text that we

---

[1] A slightly different version of this essay was published in *Monumenta Serica* 27 (1968), 208–32. The basic texts of the *Han-chi* and the *Shen-chien* used here are from the *Ssu-pu ts'ung-k'an* edition. The other editions of these works are discussed in Sections 1–2.

[2] *Hou Han-chi* (The Commercial Press: *Wan-yu wen-k'u* edition) and *Tzu-chih t'ung-chien* (Chung-hua shu-chu, 1956) entries under the tenth year of Chien-an (A.D. 205).

[3] See the prefaces in the *Ssu-pu ts'ung-k'an* edition of the *Shen-chien*.

INTRODUCTION

now have appears to be an authentic part of the original work, or at least comes from Hsün Yüeh's genuine writings. But it is quite impossible to ascertain how great a portion of the original *Shen-chien* it preserves. And there is a strong possibility that some parts of the original *Shen-chien* might have been interpolated into the "discourses" (*lun*) in the present *Han-chi*. A study of the text of the *Shen-chien* must therefore include an examination of the text and transmission of all of Hsün Yüeh's surviving writings, especially the *Han-chi*.

1. Recorded Editions of the *Han-chi*

The essential authenticity of the present *Han-chi* poses no serious problem. The work was highly prized in the Six Dynasties era (220–580),[4] and its style and genre were emulated by many later historians.[5] According to the critic Liu Chih-chi (661–721), the *Han-chi* in the pre-T'ang period was admired even more than the original *Han-shu*.[6] In the T'ang dynasty, Emperor T'ai-tsung (r. 627–649) had a very high regard for the work.[7] *Ssu-k'u ch'uan-shu tsung-mu, t'i-yao*, mentions that the *Han-chi* was made an optional subject in the civil examinations of the T'ang period, which indicates a relatively wide circulation of its text.[8]

The *Han-chi* in 30 *chüan* is recorded in the bibliographical treatises of the *Sui-shu* (comp. 629–641–656), the *T'ang-shu* (comp. 934–945; its bibliographical treatise is based on an earlier work, *Ch'ün-shu ssu-pu lu*, comp. 721), and the *Hsin T'ang-shu* (1023–1065). It is also listed in the catalogues of the Sung Imperial Libraries—*Ch'ung-wen tsung-mu* (1034–1038) and *Chung-hsing kuan-ko shu-mu* (ca. 1178–1220)—and in private bibliographies of Sung times—*T'ung-chih* (1104–

---

[4] Chang Fan's *Hou Han-chi* (fragmentary), quoted in *San-kuo chih* 10:10a.

[5] Liu Chih-chi, *Shih-t'ung* (*Ssu-pu ts'ung-k'an* ed.) 1:5.

[6] *Ibid.*, 2:3.

[7] *T'ang-shu* 62:9b; *Chen-kuan cheng-yao* (*Ssu-pu ts'ung-k'an* ed.) 2:40a.

[8] *Ssu-k'u ch'uan-shu tsung-mu* [*t'i-yao*] (Commercial Press: *Wan-yu wen-k'u* ed.) 10, p. 55.

INTRODUCTION

1162), *Sui-ch'u-t'ang shu-mu* (1127–1194), *Chün-chai tu-shu chih* (prefaced 1151), and *Chih-chai shu-lu chieh-t'i* (1234–1236).[9] Miscellaneous quotations from the work are found in the three important T'ang encyclopedias: *Pei-t'ang shu-ch'ao* (605–617), *I-wen lei-chü* (624), *Ch'u-hsüeh-chi* (ca. 725), and the T'ang annotations of the *Wen-hsüan* (ca. 658 and 718).[10]

So far four (or perhaps three) printed editions of the *Han-chi* from Sung times are known, two from the Northern and two (or perhaps one) from the Southern Sung:

(1) The Hsiang-fu (1008–1016) edition printed in Ch'ient'ang (present Chekiang). This is mentioned as the first printed edition of the *Han-chi* in Wang Chih's preface to edition no. 3 below.[11]

(2) The T'ien-sheng (1023–1032) edition printed in I-chou (present Szechwan). This is mentioned by Li T'ao (1115–1184), as quoted in the *Wen-hsien t'ung-k'ao*.[12]

(3) The Shao-hsing (1131–1160) edition printed in Chekiang. This seems to be the first combined edition of the *Ch'ien Han-chi* (*Han-chi*) and the *Hou Han-chi*. Its preface by Wang Chih is dated 1142. This edition was listed in the early Ch'ing Imperial Library Catalogue, *T'ien-lu lin-lang hou-mu* (*ch'ien-mu* comp. 1744–1775; *hou-mu* in 1797). The T'ien-lu

---

[9] *Sui-shu* 33:4b; *T'ang-shu, Chih* 26:16b; *Hsin T'ang-shu* 58:3a; *Ch'ung-wen tsung-mu* (*Yüeh-ya-t'ang ts'ung-shu* ed.) 2:6b; *Chung-hsing kuan-ko shu-mu* (fragmentary in *Ku-i shu-lu ts'ung chi*, Peiping 1933) 2:5b; *Tung-chih* (Commercial Press: *Wan-yu wen-k'u* ed.) 65, p. 772; Yu Mou, *Sui-ch'u t'ang shu-mu* (in *Shuo-fu* 28, Han-fen-lou ed.) 7b; Chao Kung-wu, *Chün-chai tu-shu chih* (Commercial Press, 1937), 2a, p. 111; Ch'en Chen-sun, *Chih-chai shu-lu chieh-t'i* (Kuang-ya shu-chu ed.) 4:18a.

[10] *Pei-t'ang shu-ch'ao* (Nan-hai K'ung-shih 1888 ed.) 3:2b, 9:2a, 13:3b, 26:4a, 33:2b, 41:5a. *I-wen lei-chü* (Wang-shih 1587 edition) 12:12a, 13, 22a and *pass*. *Ch'u-hsüeh-chi* (Huang-shih 1888 ed.) 24:25a and *pass*. *Wen-hsüan* (*Ssu-pu ts'ung-k'an* ed.) 1:1b, 3a, 6b, 29a; 2:35a; 10:27a, 29a; 20:38a; 21:4a; 26:16a; 27:22b; 37:26a; 43:23a; 47:44a.

[11] *T'ien-lu lin-lang shu-mu hou-pien* (Wang-shih 1884 ed.) 4:6. Fu Tsenghsiang, *Ts'ang-yuan ch'ün-shu t'i-chi, Hsü-chi*, 1938 ed.) 1:26.

[12] *Wen-hsien t'ung-k'ao* (Commercial Press: *Wan-yu wen-k'u* ed.), p. 1631.

INTRODUCTION

collection was later transferred to the Chao-jen-tien Palace and this edition was reported missing.[13]

(4) Another (?) Chekiang edition (which might be the same as no. 3). A facsimile manuscript of this edition has been in the possession of Ch'ü Yung, who made a detailed description of it in his *T'ieh-ch'in t'ung-chien lou ts'ang-shu mu-lu* (preface dated 1857).[14] Huang P'ei-lieh (H. Jao-p'u, 1763–1825) collated a similar facsimile with edition no. 6; the collated copy was later in the possession of Fu Tseng-hsiang, who made a second collation with edition no. 6 and published his principal findings in the *Ts'ang-yuan ch'ün-shu t'i-chi*.[15]

All of these Sung editions were described as being very corrupt. Furthermore, Cheng Ch'iao (1104–1162) in his *Tung-chih* testifies that by his time the *Han-chi* had somewhat fallen into oblivion.[16]

From the Ming period, three printed editions are recorded:

(5) Lü Nan's collated edition with preface by Ho Ching-ming. According to Sung Lo's preface in edition no. 8, this was produced sometime in the Ch'eng-hua (1465–1487) and the Hung-chih (1488–1505) periods. Mo Po-chi also attributes it to the Hung-chih period.[17] But according to the *Seikado hiseki-shi* 靜嘉堂祕籍志 (Tokyo 1917–1919) 17:19a, Ho Ching-ming's preface is dated the fifteenth year of Cheng-te (1520), and Lü Nan's the sixteenth year of Cheng-te (1521).

(6) Huang Chi-shui's edition. According to Huang's preface from 1548, the work was a facsimile reproduction of a Sung copy.[18]

---

[13] See note 12.
[14] Ch'ü-shih edition (prefaced 1857) 9:1, "13 columns on each half-page, 24 characters per column."
[15] 1:26–31a.
[16] *Tung-chih*, p. 772.
[17] Mo Po-chi, *Wu-shih-wan-chüan-lou ts'ang-shu mu-lu*, *Ch'u-pien* (1936 ed.), 4:2a, pp. 243–246.
[18] Huang's preface, in *Ssu-pu ts'ung-k'an* ed., 2b. See also K. T. Wu, "Ming Printing and Printers," *Harvard Journal of Asiatic Studies* 7 (1943), p. 242.

INTRODUCTION

(7) The Nan-chien edition, printed by the *Nan-ching Kuo-tzu chien* (Imperial Academy of the Southern Capital) in the twenty-sixth year of Wan-li (1598).[19]

The post-Ming printed editions of the *Han-chi* include:

(8) Chiang Kuo-hsiang's edition with collations (1696). According to Mao Ch'i-ling's preface, collations were made with editions nos. 6 and 7, and a certain Sung copy. The latter is, however, not mentioned in Chiang's preface to his corrigenda, the *Liang Han-chi tzu-chü i-t'ung k'ao*.[20]

(9) The Hsüeh-hai t'ang edition (1876), with corrigenda by Ch'en P'u.

(10) The *Lung-hsi ching-she ts'ung-shu* edition, by Cheng Kuo-hsün (preface to the *ts'ung-shu* dated 1917). The work is based on edition no. 8, but includes several old prefaces and notes of collation.

(11) The *Ssu-pu ts'ung-k'an* edition, a photolithographic reproduction of edition no. 6.

Another collation, *Ch'ien Han-chi chiao-shih* (in *Nan-ch'ing cha-chi*, 1894 edition), was made by Niu Yung-chien, comparing the *Han-chi* with the *Han-shu*. All these collations and corrigenda are, with the exception of Fu Tseng-hsiang's, of only minor importance; but through these records of collations, we may trace the authority of the three popular editions of the *Han-chi* (i.e., nos. 8, 10, 11) to two, or probably three, Sung copies: (a) edition no. 4 (or perhaps 3) used by Fu Tseng-hsiang; (b) a different copy followed by Huang Chi-shui; (c) probably still another copy used for Chiang's edition.

2. RECORDED EDITIONS OF THE *Shen-chien*

The problem of the *Shen-chien* text is more complicated. The work in five *chüan* is recorded in the *Sui-shu*, the *T'ang-shu*, the

---

[19] Mo Po-chi, *Wu-shih wan-chüan-lou ch'ün-shu pa-wen* (1948 ed.), p. 119. *Tseng-ting Ssu-k'u chien-ming mu-lu* (Chung-hua shu-chu, 1959 ed.), p. 208.

[20] See also Fu Tseng-hsiang, 1:26. This edition was strongly recommended by the *Tseng-ting Ssu-k'u chien-ming mu-lu* p. 208.

INTRODUCTION

Hsin T'ang-shu, the *Chung-hsing kuan-ko shu-mu*, the *T'ung-chih*, the *Sui-ch'u t'ang shu-mu*, the *Chih-chai shu-lu chieh-t'i*, the *Wen-hsien t'ung-k'ao*, and many bibliographical works of Ming and Ch'ing times.²¹ Consequently, its textual problems have been overlooked by many serious critics, even by such perceptive scholars as the editors of the *Ssu-k'u ch'üan-shu* and by Chang Hsin-cheng in his *Wei-shu t'ung-k'ao*.

The *Shen-chien* is not, however, listed in the fragmentary *Ch'ung-wen tsung-mu* (comp. 1034–1038, recollected in 1799), nor in the *Chün-chai tu-shu-chih* (comp. 1151; editions of 1250 and 1249), nor in the *I-wen-chih* of the *Sung-shih* (comp. 1343–1345, its bibliographical treatise is based on a number of Sung official catalogues). The listing of the *Shen-chien* in five *chüan* under the section of "Miscellaneous Writings of Individual Authors" (*pieh-chi* 別集), in Chao Hsi-pien's supplement (*Fu-chih*) to the *Chün-chai tu-shu-chih* (prefaced 1250) only adds to the confusion.²²

The earliest existing record of a Sung printed edition of the *Shen-chien* is a postscript (*t'i-tz'u* 題辭) by Yu Mou dated 1182.²³ This edition had been in the possession of the Sung scholar Huang Chen (fl. 1256–1270), who made a detailed description of it, criticizing it severely.²⁴ Both Yu Mou's postscript and Huang Chen's remarks will be discussed in Section 3.

A facsimile of a Sung copy is mentioned in Sun Ts'ung-t'ien's *Shang-shan-t'ang shu-mu*.²⁵ A Yuan edition printed by Ch'en Tzu-jen is recorded in the [*Tseng-ting*] *Ssu-k'u chien-ming mu-lu* (prefaced 1908).²⁶ An edition designated *Hsiao Hsun-tzu* 小荀子 is mentioned among six other philosophical works (*tzu*) in Chi

---

²¹ *Sui-shu* 34:1; *T'ang-shu, Chih* 27:1b; *Hsin T'ang-shu* 59:1b; *Chung-hsing kuan-ko shu-mu* 4:5a; *T'ung-chih* 66, p. 785; *Sui-ch'u-t'ang shu-mu* 21a; *Chih-chai shu-lu chieh-t'i* 9:6; *Wen-hsien t'ung-k'ao* 209, p. 1720; *Wen-yuan-ko shu-mu* (comp. 1441; in *Tu-hua chai ts'ung-shu*) 7:5b; *Nei-ko ts'ang-shu mu-lu* (comp. 1605; in *Shih-yuan ts'ung-shu*) 2:41b.

²² *Chün-chai tu-shu-chih, Fu chih*, pp. 628–629.

²³ The *T'i-tz'u* was incorporated in the *Hsiao Hsün-tzu* of the *Tzu-hui* edition.

²⁴ Huang Chen, *Tz'u-ch'i Huang-shih jih-ch'ao fen-lei* (Tz'u-ch'i Feng-shih keng-yü-lou ed.) 57:6b–7a.

²⁵ Ch'en shih 1929 ed., 13b.

²⁶ *Tseng-ting Ssu-k'u chien-ming mu-lu* 9, p. 382.

INTRODUCTION

Chen-i's *Chi Ts'ang-wei ts'ang-shu-mu*, which generally lists only Sung and Yuan editions.[27]

Several editions of the *Shen-chien* were printed in Ming and post-Ming times:

(1) Li Lien's edition; editor's preface dated 1518.[28]

(2) Huang Hsing-tseng's annotated edition. This is the only annotated edition of the *Shen-chien*. Editor's preface is dated 1519; Wang Ao's preface, also 1519; Ho Yuan-fu's preface, 1525; Ch'iao Yü's postscript, 1521. Huang's collations indicate that he made use of an earlier copy than that on which edition no. 4 was based.

(3) Chang Wei-shu's edition printed in 1533 (?).[29]

(4) The *Hsiao-Hsün-tzu* in the *Tzu-hui* edition. This was printed during the period 1522–1566 and reprinted in 1577;[30] a photolithographic reprint was made by the *Han-fen-lou* (the Commercial Press). The *Tzu-hui* collection consists of twenty-four philosophic works (*tzu*), of which the *Hsiao Hsün-tzu* is the sixth. Similar collections of philosophic works had been printed in Sung times. The *Hsiao Hsün-tzu* was not included in the *Liu-tzu* 六子 collection printed in the Sung dynasty; but it was in the *Shih-erh-tzu* 十二子 collection which shows many corruptions of Sung taboo words, suggesting a Sung origin; and another *Hsiao Hsün-tzu*, probably also a Sung edition, has been mentioned before.[31] Since the present *Hsiao Hsün-tzu* in the *Tzu-hui* incorporates both Yu Mou's postscript and Li Lien's preface, it may well be a reproduction of a Sung copy with collations from edition no. 1.

(5) The *Liang-ching i-pien* edition. This was printed by Hu

---

[27] In *Shih-li-chü ts'ung-shu*, 1805 ed., 22b.

[28] The preface is incorporated in the *Tzu-hui* edition.

[29] *Tseng-ting Ssu-k'u chien-ming mu-lu*, p. 382. A reprint of edition no. 2 ?

[30] Yang Shou-ching, *Ts'ung-shu chü-yao* (I-ch'iu kuan ed.) 9:10a–11b.

[31] *Tseng-ting Ssu-k'u chien-ming mu-lu*, p. 384; see also note 27. Another *Hsiao Hsün-tzu*, in the *Chu-tzu hui-han* (subtitle: *Chu-tzu p'ing-lin*), printed in 1625 by book merchants and ascribed to the editorship of Kuei Yu-kuang (1506–1571), incorporated nine selected passages from the *Shen-chien*, which must be distinguished from the complete *Hsiao Hsün-tzu* edition.

INTRODUCTION

Wei-hsin, with preface dated 1582. The text of the *Shen-chien* appears to be a perfect facsimile of edition no. 2.[32]

(6) The *Han-Wei ts'ung-shu* editions:

(a) Serial A, compiled by Ch'eng Jung in 1592. The *Shen-chien*, in *ts'e* 50–51, is based on edition no. 2.

(b) Serial B, compiled by Ho Yün-chung. The *Shen-chien* in *ts'e* 31 is a much inferior reproduction of the Serial A edition, omitting many of Huang's annotations.

(c) Serial C, compiled by T'u Lung and printed by Wang Mu in 1791, with proofreadings by Wu Tao-ch'uan. The *Shen-chien*, in *ts'e* 46, is based on that of Serial B.

(d) Serial D: the *Wang-shih chin-hsiang* edition, *ts'e* 43, printed in 1880.

(7) The *Hsiao-wan-chüan lou ts'ung-shu* edition, with collations and *Cha-chi* 札記, by Ch'ien P'ei-ming, printed in 1852.

(8) The *Tzu-shu po-chung* edition, printed by the Ch'ung-wen shu-chu 1875, based on edition no. 6, Serial B.

(9) The *Lung-hsi ching-she ts'ung-shu* edition, printed in 1917, based on edition no. 2, with Lu Wen-ch'ao's (1717–1795) "Notes of Collation" (from *Ch'ün-shu shih-pu*) and a number of *Shen-chien* quotations from the *Ch'ün-shu chih-yao* added as an appendix.

(10) The *Chung-hua shu-chü* edition printed in 1927 and reprinted in the *Ssu-pu pei-yao* in 1930. The *Shen-chien* text is based on edition no. 6, Serial C.

(11) The *Ssu-pu ts'ung-k'an* edition, a photolithographic reproduction of edition no. 2 by the Commercial Press in 1929.

Edition no. 2 is the earliest extant edition with annotations, and it had the widest circulation in later reproductions or reprints. In Huang Hsing-tseng's annotations, different readings of the *Shen-chien* quotations in the *Hou Han-shu* are noted; collations are also made with at least one other copy, which appears to be none other than an earlier copy of the *Hsiao Hsün-tzu* in edition no. 4, the one that came closest to a Sung edition.

---

[32] See also K. T. Wu (note 18), p. 249.

INTRODUCTION

On the other hand, edition no. 7 represents the best result of later textual collations. In a postscript to this edition (1852), Ch'ien P'ei-ming writes:

"The *Shen-chien* in 5 *chüan* has always been thought to be without [serious] omissions. But when I collated it with the quotations in the *Ch'ün-shu chih-yao*, I found many corruptions and omissions. In one place, there was an omission of one hundred and seventy-odd characters.[33]

"At present the most popular edition is that of Ch'eng Jung's *Han-Wei ts'ung-shu*, which is based on Huang Hsing-tseng's annotated edition of 1519–1521. Huang occasionally quotes the *Hou Han-shu* for collation, but there are inaccuracies. The academician Lu Wen-ch'ao (1717–1795) has made some dozens of corrections in his *Ch'ün-shu shih-pu*.[34] But since Lu did not see the *Ch'ün-shu chih-yao*, some of his corrections appear to be guesswork.

"In the present edition I have used Huang's annotated edition as the basic text, emending its errors and lacunae with the help of the *Hou Han-shu*, the *Ch'ün-shu chih-yao*, the *Pei-t'ang shu-ch'ao*, the *T'ai-p'ing yü-lan*, and Lu's *Ch'ün-shu shih-pu*. I have recorded all my corrections in the *Cha-chi*. In the places where a correct reading cannot be found I have kept the original version. Although Huang's annotations sometimes come close to the author's meaning, they hardly make any telling points. Therefore I have removed all of them from the present edition."[35]

Later, Sun I-jang (1848–1908) made two more corrections in his *Cha-i* (1894 edition). Liu Shih-p'ei (1884–1919) wrote a supplement to edition no. 7, in which he collated the present text with quotations from Ma Tsung's *I-lin*.[36]

---

[33] In *Ch'ün-shu chih-yao* (*Ssu-pu ts'ung-k'an* ed.) 46:8b9–9a9, and *Shen-chien* 4:12b6 respectively.

[34] "Shen-chien chiao-cheng" in *Ch'ün-shu shih-pu*, Ch'u-pien (*Pao-ching-t'ang ts'ung-shu* ed.).

[35] Ch'ien's postscript in edition no. 7.

[36] In Liu's *Tso-an chi* (*Liu Shen-shu hsien-sheng i-shu* ed.).

INTRODUCTION

3. The *Shen-chien* Text in Sung and Pre-Sung Times

In the preceding section, it was mentioned that the history of the transmission of the *Shen-chien* was somewhat confused during the Sung dynasty. In his postscript (*t'i-tz'u*) to the earliest known edition (1182) of the work, Yu Mou writes:

"Hsün Yüeh's writing 荀悅書 in 5 *chüan*. From its contents we can see that he had the intention of serving the state. The *Han-chi*, which he also wrote, contains an embryo of the present work (嘗載其略). Fan Yeh also produced an excerpt of this work in the Biography of Hsün Yüeh in the *Hou Han-shu*. The *Han-chi* has already been set in print in the Kuai-ch'i Commandery. But few people ever saw a complete copy of the present work (the *Shen-chien*). My family happened to possess such a copy. I therefore had it printed, depositing the wood blocks in the office of the Grain Transport of Chiang-hsi. However, the text of the original was marred by many omissions and mistakes. I did not dare to make any additions or omissions by my own judgment. Wherever there were doubts I left them to the enlightened."[37]

It is difficult to judge how much credit may be given to Yu Mou's effort to preserve the original *Shen-chien* manuscript. In the above postscript, he vaguely refers to the work as "Hsün Yüeh's writing," although the mention of the excerpt in the *Hou Han-shu* unmistakably refers to the *Shen-chien*. Yu Mou's reserved statement seems to have a bearing on the modest title *Hsiao Hsün-tzu* (Philosophic Writings of the Junior Master Hsün) in edition no. 4 and its predecessors. Huang Chen, who possessed a copy of Yu Mou's edition, writes:

"The *Shen-chien* in five *p'ien* (chapters) is the writing of Hsün Yüeh of the Eastern Han dynasty.... In general, the book is wordy and inconsequential 辭繁理寡; its style is also inconsistent. Ch. 1, Essence of Government 政體, and Ch. 2, Current Affairs 時事, both contain many subheadings, and quite resemble the style of the *Chou-shu* from the tomb of the Chi district 汲冢

[37] See note 23.

INTRODUCTION

(unearthed in A.D. 281); Ch. 3, Common Superstitions 俗嫌 and Ch. 4 and 5, Miscellaneous Dialogues 雜言 are mostly in the question-and-answer style, resembling Yang Hsiung's *Fa-yen*."

(Here follows a critical review of several passages of the *Shen-chien*, corresponding to 1:11a4–5, 11b2–12a5, 13a5–13b3; 3:1a4–5, 3a6–9, 3b5–9, 5b5–6; 4:5b7–9, 6b6–9; and 1:11b2–12a5 in the *Ssu-pu ts'ung-k'an* edition.)

"... His discussions on human nature (*hsing*) and feelings (*ch'ing*) encompass several sections, but very few of them make good points. The style is also very weak 卑弱, and unlike that of the *Han-chi*. I do not know whether this work is actually from Hsün Yüeh's pen. The present edition was printed by Yu Mou in the ninth year of Shun-hsi (1182)."[38]

Since Huang Chen's description and citations of the *Shen-chien* accord with all existing editions of the work (nos. 2, 4, 5, 6, 7, 8, 9, 10, and 11) which directly or indirectly followed Yu Mou's edition, his suspicion bears on all extant editions of the *Shen-chien*.

In his postscript, Yu Mou notes that seeds of the *Shen-chien* were contained in the *Han-chi*. This is the first important observation of the textual relation between these two works and thus deserves close scrutiny. In the present *Shen-chien* text there are many passages which continue or repeat Hsün Yüeh's discourses (*lun*) in the *Han-chi*, as the following table shows:

Shen-chien	*Han-chi lun*
(*Ssu-pu ts'ung-k'an* ed.)	(*Ssu-pu ts'ung-k'an* ed.)
1:4a6–8	10:4
1:4b4–6	10:4
Quotation in *Ch'un-shu chih-yao* 46:2b3–5	10:4
2:2a7–2b4	10:4
2:15a7–15b1	17:10a
2:6a8–7a3	23:10b–11
2:5a8–5b5	28:5a–6a

[38] See note 25.

INTRODUCTION

The above passages in the *Shen-chien* contain almost verbatim repetitions of the corresponding *Han-chi* discourses and pose serious textual problems. Since Yu Mou's edition of the *Shen-chien* was produced during a time when the record of the transmission of the work was quite confused, the question may be asked: Would the so-called rare copy of the *Shen-chien* in Yu's possession be a scissors-and-paste reproduction from the various works expressly referred to in his postscript?

To answer this question, I have collected the fragmentary citations of the *Shen-chien* in the various pre-Sung authorities and compared them with the present *Shen-chien* text. In the following table the asterisks in the third column indicate the degree of discrepancy between the pre-Sung citation and the present text of the *Shen-chien*: * minor, ** moderate, *** important.

Pre-Sung Authorities	Present *Shen-chien*	Discrepancies
Hou Han-chi		
29:14b10–15a1	1:1a4–6	**
29:15a1–16b1	1:2b4–6b1	*
29:16b1–6	2:15b5–16a5	* (** in last 3 sentences)
Hon Han-shu		
62:11a5–8	1:1a4–6	*
62:11a8–14a1	1:2b4–6b1	**
62:14a1–4	2:15a7–15b2	*
62:14a4–14b3	2:15b5–16b5	* (*** in last 3 sentences)
I-lin (preface of 786 by Tai Shu-lun)[39]		
5:2a3–4 (abstract)	1:1a4–7	
5:2a5–2b1 (abstract)	1:7a7–8a4	
5:2b2	1:12b4–5	
5:2b3–4 (fragment)	2:1a6–1b3	

[39] *Ssu-pu ts'ung-k'an* ed. The work, by Ma Tsung, was based on Yü Chung-yung's *Tzu-shu ch'ao* (ca. 475–548).

INTRODUCTION

5:2b5	2:13b8–9	
5:2b6	3:4b2–3	***
5:2b7	4:1a8	***
5:2b8	4:6a9–6b1	
5:2b9–3a2	5:9b6–10a5	*

Pei-t'ang shu-ch'ao
  (comp. 605–617)

27:2a	2:2a4–5	(order reversed)
55:3b	2:15b5–16a3	**
90:8a	3:2b8–3a1	
136:3a	4:1a8–1b2	***

Ch'ün-shu chih-yao
  (comp. 628–630)

46:1a5–7	1:1a4–7	**
46:1a7–8	1:2a8–9	**
46:1a8–3b6	1:2b1–6b1	**
46:3b7–4b8	1:6b6–8a4	*
46:4b9–5a2	1:9a5–8	
46:5a3–4	1:10a3–6	
46:5a5–9	1:14a1–9	
46:5b1–5	1:14b8–15a3	
46:5b5–7	1:10b5–8	**
46:5b8–6a7	2:2a2–2b4	*
46:6a8–6b8	2:15a7–16a5	*
46:6b9–7a3	4:1a8–1b3	***
46:7a4–8a5	4:2a5–4a6	
46:8a6–8b2	4:5b1–7	*
46:8b3–8	4:6a3–9	
46:8b9–9b2	4:12b–8	(171 characters missing from the *Shen-chien*)
46:9b2–5	5:2b9–3a3	
46:9b5–7	5:8a8–8b2	*

Wen-hsüan Annotations
  文選注 (by Li Shan
  and Five High Officials
  五臣 ; memorial dated 718)

INTRODUCTION

10:4b	4:8a7
10:9b	4:6b5
28:3b	4:1b2
36:19b	2:8b3–4
37:30a	4:8a7–8
37:32a	1:14a6
43:14a	4:8a7
57:12a	4:8a7
58:27a	1:14a6
58:32a	4:8a7

Distribution of pre-Sung quotations in the
Ssu-pu ts'ung-k'an edition of the *Shen-chien*

Chüan Nos.		1	2	3	4	5
Pages nos.	1	a	ab		ab	
	2	ab	ab	b	ab	b
	3	ab		a	ab	a
	4	ab		b	a	
	5	ab			b	
	6	ab			ab	
	7	ab				
	8	a	b		a	ab
	9	a				b
	10	ab				a
	11					
	12	b			b	
	13		b			
	14	ab				
	15		ab			
	16		a			
Total pages		15	16	8	13	10

From this analysis, we can see that the quotations of the *Shen-chien* in the pre-Sung works cover a fairly great number of pages in the present *Shen-chien* text, which indicates that the present *Shen-chien* is not totally a Sung forgery. On the other hand, the marked discrepancies between these quotations and

INTRODUCTION

the present *Shen-chien* text also show that the present *Shen-chien* is not a mere scissors-and-paste collection of earlier quotations. Of the six *Shen-chien* passages that contain verbatim repetitions of the *Han-chi lun* as indicated above, four (*Shen-chien* 1:4a6–8; 1:1b4–6 and *Ch'ün-shu chih-yao* 46:2b3–5; *Shen-chien* 2:2a7–2b4; 2:15a7–15b1) can be warranted by the pre-Sung quotations, which also proves the innocence as well as the insight of Yu Mou's statement of the textual relation between the present *Shen-chien* and *Han-chi*.

4. Problems of the Sung or Pre-Sung *Han-chi* Text

From the previous discussion, we may safely conclude that the present text of the *Shen-chien*, although quite corrupt, is not a Sung forgery. The question may be asked: Would those identical passages in the *Shen-chien* and the *Han-chi* pointed out in the preceding section discredit the latter instead of the former? For, as mentioned before, although the *Han-chi* had a wide circulation in T'ang times, the work had fallen into obscurity under the Sung dynasty and all of its Sung editions were reported to be very corrupt.

In the present *Po-na* edition of the *Han-shu* there are a number of notes—most of them indicating the tabooed personal names and the styles of the Han emperors—which bear the introductory phrase "Hsün Yüeh says."[40] In Yü Ching's memorial upon the submission of the Sung-collated edition of the *Han-shu*, dated the second year of Ching-yu (1034), these are mentioned as quotations from Hsün Yüeh's *Han-chi* added to the *Han-shu* by some previous annotators of the latter work.[41] However, none of these quotations can be found in the present *Han-chi* or in the various records of collation discussed in Section 1. These materials must have been deleted from the Sung editions of the *Han-chi* sometime

---

[40] *Han-shu* 1A:1a3; 2:1a3; 3:1a3; 4:1a3; 5:1a3; 6:1a3; 7:1a3; 8:1a3; 9:1a3; 10:1a3.

[41] See Yü Ching's memorial appended to the *Po-na* edition of the *Han-shu*, 1b.

between 1034 (the year Yü Ching submitted his memorial) and 1142 (the year Wang Chih wrote his preface to edition no. 3 of the *Han-chi*, which is one of the earliest editions of the present *Han-chi*). This, together with the statements by Cheng Ch'iao, Wang Chih, and Li T'ao referred to in Section 1, indicates that the text of the *Han-chi* was severely tampered with in Sung times when neo-Confucianism made extensive inroads into Chinese historiography.

Some important interpolations in the *Han-chi* appear as "discourses" inserted by Hsün Yüeh. All the passages identical with the *Shen-chien* are to be found in these "discourses". In his memorial to the Han emperor upon the submission of the *Han-chi*, Hsün Yüeh clearly differentiated between such "discourses" (*lun*) and the text:

"In the *Han-chi*, when a passage is introduced by the words *nien* 年 (year), *pen-chi* 本紀 (annal), *chih* 志 (treatise), and *chuan* 傳 (biography), the material comes from the *Han-shu* text 書家本語; when it is introduced by the word *lun* 論 (discourse), it is your minister Yüeh's personal comment which gives a summary evaluation of an important event."[42]

However, in the present *Han-chi* Hsün Yüeh's discourses are introduced not by the word *lun* but by the phrase "Hsün Yüeh says 荀悅曰." This is obviously a later alteration. For even if Hsün Yüeh had changed his manner of presenting a discourse, the introductory phrase would have read *"ch'en Yüeh yüeh* 臣悅曰" (your minister Yüeh says), which was the proper phrase of reference to the author himself in the *Han-chi*.[43]

Clear evidence that the original discourses in the *Han-chi* actually correspond to Hsün Yüeh's description is found in *Han-chi* 28:5b. The present *Han-chi* text reads: "In the second year of Chien-p'ing (4 B.C.), the fourth month, *wu-wu* day, the *Ta-ssu-k'ung* Chu Po became the *Yü-shih ta-fu* (Grand Overseer). The discourse (*lun*) reads...." The discourse that follows gives a brief comment on the establishment

---

[42] *Han-chi* 30:26b.
[43] *Han-chi* 1:1a; 25:5a; 30:26b. For further discussions, see Section 5.

INTRODUCTION

and subsequent changes of the office of the *Ch'eng-hsiang* (Lieutenant Chancellor) and *San-kung* (the three Ducal Ministers, i.e., the *Ssu-t'u*, the *Ssu-ma*, and the *Ssu-k'ung*, which replaced the *Ch'eng-hsiang*, the *T'ai-wei*, and the *Yü-shih ta-fu*). This discourse is no doubt a typical *lun* in the *Han-chi* as described by Hsün Yüeh himself in his memorial quoted above. Since the discourse follows the entry noting Chu Po's appointment as *Yü-shih ta-fu*, it may easily be mistaken as Chu Po's own comment on the event. This, however, is not possible, because immediately before this passage Chu Po is recorded as having presented a memorial to the throne in which he recommended the abolition of the *San-kung* and the reinstitution of the *Ch'eng-hsiang* and the *Yü-shih ta-fu*; consequently, Chu Po could not have made this discourse, which censures the change and upholds the *San-kung* as the proper institution. That the introductory word *lun* was left intact in this unique case seems to have been due to the fact that the editor who changed the word *lun* into the phrase "Hsün Yüeh says" in the *Han-chi* had mistaken the discourse as Chu Po's counsel.

Still more important than these alterations in phraseology are the possible changes of the content of some of the discourses introduced by the phrase "Hsün Yüeh says" in the present *Han-chi*. For the phrase "Hsün Yüeh says" logically covers a broader range of Hsün Yüeh's writings than the original *lun* as defined by Hsün Yüeh himself. Later editors of the *Han-chi* might have introduced some of Hsün Yüeh's other writings to elucidate the summary *lun* discourses in the *Han-chi* after changing the restrictive term *lun* to the broad term "Hsün Yüeh says."

With regard to this question, we will look at the abovementioned *lun* on the establishment of the *Ch'eng-hsiang* and *San-kung*. The *lun* reads:

"The office of the *Ch'eng-hsiang* and the *San-kung* has been changed several times [during the Han dynasty]. This is not in good accord with the canons. Originally, the office of the *Ch'eng-hsiang* was established in the state of Ch'in. The Ch'in, being a feudal state of the second grade, had two ministers appointed [by the Chou king]. Hence, there were the Left and the Right *Ch'eng-hsiang*, and not the *San-kung*.

"The Ode reads: 'Never idle, day or night, in the service of the One Man.'[44] The 'One Man' means the Son of Heaven; under him there should be an office of the Triumvirs. This being described in the *I-ching* as the three legs of a tripod,[45] means the *San-kung*, who should be in charge of the tripartite government. Therefore, the office of the *San-kung* is a better institution."[46]

This discourse, consisting of eighty-five Chinese characters, accords well with the criterion of a *lun* as that "which gives a summary evaluation of an important event," as specified in Hsün Yüeh's memorial. If we accept this as an exemplar of Hsün Yüeh's original *Han-chi lun*, then the following discourses introduced by the phrase "Hsün Yüeh says" in the present *Han-chi* would fit the same criterion with respect to length and content: *Han-chi* 3:12b; 4:2a; 5:4b; 7:5b; 8:16b; 9:9a; 9:16b; 9:17a; 12:4b; 15:5; 17:10a; 20:5b; 27:12a.

On the other hand, the remaining discourses in the present *Han-chi* are all several hundred characters long. They cover topics of general significance, and some of them can be detached from the rest of the *Han-chi* to form independent essays of themselves. Furthermore, a number of these discourses can be divided into two parts: the first being a short comment directly relevant to the corresponding *Han-chi* notation of an event, the second consisting of a general discussion only loosely related to the first part.

For instance, the discourse in *Han-chi* 16:13–14 contains two parts: 16:13a7–8 and 13b8–14b7. The first section is a brief comment (29 characters) on the dethronement of the Prince of Ch'ang-i in 86 B.C., which fits in well with our criterion for

---

[44] See James Legge, tr., *The Chinese Classics* (Hong Kong, 1960), IV, p. 543.
[45] *Chou-i cheng-i* (in the *Shih-san-ching chu-shu*, I-wen yin-shu-kuan photo-lithographic edition) 5:22b.
[46] *Han-chi* 28:5b.

INTRODUCTION

a *Han-chi lun*. The second section is a lengthy discussion (635 characters) of six types of rulers and six comparable types of ministers, and is only indirectly relevant to the fact of the prince's dethronement. This second part is introduced by the phrase "*ku-yüeh* 故曰" (therefore, it is said), which is logically and grammatically separate from the first part and fits in with the broad phrase "Hsün Yüeh says."

A similar case may be found in the discourse in *Han-chi* 28:7. This *lun* again contains two parts: the first (7a8–9) consists of 20 characters commenting on Li Hsin's censure of the emperor's wet-nurses; the second, beginning with a quotation from Confucius, consists of an independent essay of more than 200 characters expounding the author's strictures on petty men who received improper imperial favors, and is only indirectly related to the incident recorded. Similar divisions may be made in two other discourses in *Han-chi* 8:8b–9a (8b4–7 and 8b7–9a), and *Han-chi* 25:6b–7a (6b4–7 and 6b8–7a).

A more obvious case is the discourse in *Han-chi* 23:9–11. The passage in the middle of this discourse (23:10b7–9) records an unfavorable comment made by Emperor Hsüan about his heir-apparent's devotion to Confucianism. Since a similar comment is recorded in the "Annals of Emperor Yüan," *Han-shu* 9:1, the *Han-chi* passage may belong to the category of "materials taken from the *Han-shu* text." The section preceding this passage in the *Han-chi* discourse contains a review of the achievements of the various Han rulers up to the reign of Emperor Hsüan, whereas the section following the passage presents a general discussion on Legalism and Confucianism. These three sections (two discourses and one historical entry) are combined into one unit introduced by the phrase "Hsün Yüeh says" in the present *Han-chi*.

In *Tzu-chih t'ung-chien*, Ssu-ma Kuang uses the phrase "Hsün Yüeh says" to introduce quotations from Hsün Yüeh's discourses on history.[47] Since Ssu-ma Kuang quotes the *Han-chi*

---

[47] Chung-hua shu-chu, 1956 ed., p. 333 *et pass*. The phrase actually is "*Hsün Yüeh lun-yüeh*" (Hsün Yüeh's discourse reads).

INTRODUCTION

discourses in a flexible and selective manner,[48] it is difficult to tell whether he initiated the practice of using this phrase or merely followed some earlier editors of the *Han-chi* in Sung or pre-Sung times. In view of the high prestige enjoyed by Hsün Yüeh in the T'ang and pre-T'ang periods, it is quite possible that some of his concise *lun* in the *Han-chi* were expanded or elucidated by the editors of those times who reproduced, for scholastic or other purposes, some materials from Hsün Yüeh's other writings (such as the *Shen-chien*) as annotations to the *Han-chi lun*. These added materials may have been subsequently merged into the *Han-chi lun* proper when the corrupt *Han-chi* manuscripts underwent reeditions in later periods. This might also partly account for the utter corruption of the post T'ang *Shen-chien* manuscript—its contents being extracted to provide annotations to the *Han-chi*.[49] Unfortunately, for lack of other substantial evidence, the investigation of this point can go no further.

5. DATES OF THE *Han-chi* AND THE *Han-chi Hsü*

Another important question, though only indirectly related to the textual problems of Hsün Yüeh's writings, remains to be answered—namely, the dates of the *Han-chi* and the *Shen-chien*. The biography of Hsün Yüeh in the *Hou Han-shu* mentions only that Hsün Yüeh wrote the two works during the Han-Wei transition period (196–219), and gives no specific dates. In

---

[48] For instance, compare *Han-chi* 4:2a, 10:2b–4b and 20:11 with *Tzu-chih t'ung-chien*, pp. 385, 607–608, 886.

[49] Two annotated editions of the *Han-chi* are listed in *Hsin T'ang-shu* 58:3a, i.e., Ying Shao's *Han-chi chu* and Ts'ui Hao's *Han-chi yin-i*. The former is not mentioned in *Sui-shu* 33, nor in *T'ang-shu* 46. The latter is listed in *T'ang-shu* 46:16b, but not in *Sui-shu* 33. Yao Chen-tsung, in his *Hou-Han i-wen-chih* (in *Erh-shih-wu-shih pu-pien*, K'ai-ming shu-tien ed.), pp. 47–48, suspects Ying Shao's annotations to be a later attribution. Ts'ui Hao's work is mentioned in the *T'ung chih* 65, p. 772, but not in the *Wen-hsien t'ung-k'ao*; whereas Ying Shao's work is not mentioned in either of the two works. All these annotations of the *Han-chi* seem to have been lost in Sung times. Some of them might have been merged into the *Han-chi* proper.

INTRODUCTION

Yuan Hung's (328–376) *Hou Han-chi* (Chronicles of the Later Han), under the tenth year of the Chien-an (205), an entry reads:

"The eighth month: the *Shih-chung* (Attendant at the Palace) Hsün Yüeh wrote [a work] on the good and bad points of state affairs and designated it *Shen-chien*, which was now completed and submitted to the throne."

There follows an excerpt from the *Shen-chien*.

"... Yüeh, styled Chung-yü, was a native of Ying-ch'uan. When he was still young, he was already marked for his talent in reasoning. He studied both the Confucian canons and history. By this time Lord Ts'ao [Ts'ao] had usurped the state power, while the emperor only solemnly folded his hands. The emperor was interested in literature and was talented in letters. He considered the *Han-shu* too voluminous and ordered Yüeh to simplify it; [the result was] the *Han-chi* in thirty chapters."[50]

The above entry constitutes an important source of biographical information. It should be noted that in this entry the principal event is the submission of the *Shen-chien*, whereas the second paragraph is but a supplement providing additional facts about the author and his other activities—activities that might have occurred before or after the principal event. The same sequence of narration was adopted by Fan Yeh (398–446) in his biography of Hsün Yüeh in the *Hou Han-shu*.[51] This has led some to the erroneous impression that Hsün Yüeh first wrote the *Shen-chien*, which was highly prized by the Han emperor, who then commissioned him to compile the *Han-chi*.[52] Such a view was not expressed by Ssu-ma Kuang in his *Tzu-chih t'ung-chien*. There he describes the *Shen-chien* in a manner similar to the way Yuan Hung refers to the work, except that he does not mention the *Han-chi* compilation.[53]

The career of Hsün Yüeh during the first five years of the

---

[50] Yuan Hung, *Hou Han-chi* 29, pp. 352–354.

[51] *Hou Han-shu* 62:11–14.

[52] For instance, see Yao Chen-tsung, *Sui-shu ching-chi-chih k'ao-cheng* (in *Erh-shih-wu-shih pu-pien*), p. 220.

[53] *Tzu-chih t'ung-chien*, pp. 2064–2065.

INTRODUCTION

Chien-an period (196–200), especially his compilation of the *Han-chi*, is recounted in detail in a *"Han-chi hsü"* (*Han-chi* Preface) in the present *Ssu-pu ts'ung-k'an* edition of the *Han-chi*. This preface (hereafter cited as Preface A) is attributed to Hsün Yüeh. But it differs completely from the passage quoted as "preface" of the *Han-chi* in the Biography of Hsün Yüeh (hereafter cited as Preface B).[54] Consequently, we now have two different prefaces of the *Han-chi* ascribed to Hsün Yüeh.

Wang Hsien-ch'ien, in his *Hou Han-shu chi-chieh*, accepted Preface A as Hsün Yüeh's original preface to the *Han-chi* and considered Preface B as the beginning section of the *Han-chi* proper.[55] This is obviously a misstatement. For Preface B consists of not one but two (or possibly four) excerpts from the beginning and the concluding sections of the *Han-chi* in the present *Ssu-pu ts'ung-k'an* edition (Preface B—*Han-chi* 1:1a2–8; 30:26b5–6; 1:1a8–9; 30:26b6–9).

Preface B, especially its concluding part which in the present *Han-chi* takes the form of a memorial upon the submission of the *Han-chi* to the throne, appears in style and content to be the original writing of Hsün Yüeh. Preface A differs from Preface B in many respects. In Preface B, Hsün Yüeh twice refers to himself as *"ch'en* Yüeh (your minister Yüeh); only at the end of the memorial does he give his signature as *"Shih-chung* Yüeh." In Preface A, on the other hand, the author of the *Han-chi* is throughout referred to as "Yüeh" or "Hsün Yüeh."

The tone of the two prefaces also differs remarkably. Although both Preface A and B are given to magnifying the virtue of the Han dynasty and the importance of historical records, the tone in Preface B appears to be considerably more reserved and modest when referring to the personal career of Hsün Yüeh. Whence the expressions: "I was acting as the *Mi-shu-chien* merely to fill the vacancy"; "humbly I received the enlightening decree"; "secretly I considered"; "cautiously and sincerely I simplified"; "hoping that I will leave few things

---

[54] *Hou Han-shu* 62:14b–15.
[55] *Hou Han-shu chi-chieh* (Wang shih 1915 edition).

INTRODUCTION

out"; "not being able to attain the golden mean"; "if there are some good and bad points in it, I wait for the judgment of the superior man";[56] and "crudely presenting the major events"; "I, the petty minister, sincerely write."[57] All these are in sharp contrast to the grandiloquence of Preface A. Only on one occasion in Preface A is the author of the *Han-chi* referred to modestly, namely in the phrase, "although [it is] said that the author is unrefined and shallow"[58]—a wording which is, however, somewhat impersonal in tone. In Preface A there is also the strange sentence, "the *Chi* reads 記曰: this four hundred and sixteenth year, meaning that the work was submitted to the throne in the Ken-ch'en year (A.D. 200)."[59] This sentence is completely out of place in Preface A and resembles an explanatory note to a specific statement in Preface B.[60]

However, the eulogistic expressions in Preface A, in which the Han rulers are referred to as "Ancestors" (*tsu* 祖 and *tsung* 宗) and "our late emperors" (*hsien-ti* 先帝), do tend to show that it was written in the Han period. In form the document resembles an official record kept in the Han court, commemorating the completion of the *Han-chi* and recounting the events which led to its commission.

In Chiang's edition of the *Han-chi* (1696), Preface A is appended to the *Han-chi mu-lu* (Table of Contents). It was subsequently taken by Yao Chen-tsung as part of the *Han-chi mu-lu* itself. In his *Sui-shu ching-chi-chih k'ao-cheng*, Yao wrote:

"This document appears at the end of the *Han-chi mu-lu*. Its

---

[56] *Han-chi* 1:1.
[57] *Han-chi* 30:26b, 27a.
[58] *Han-chi hsü*, 2b.
[59] *Han-chi hsü*, 2a.
[60] The statement appears in *Han-chi* 30:26b, which reads: "This four hundred and twenty-sixth year of Han." It is quoted as "this four hundred and sixth year of Han" in the Biography of Hsün Yüeh, *Hou Han-shu* 62:15a. The discrepancies are due to successive corruptions. The explanatory note is needed because the *Han-chi* covers only "two hundred and forty years of the Former Han, including Wang Man's reign" as specified in Preface A (*Han-chi hsü*, 1a), and the *Keng-sheng* year (A.D. 200) is only the one hundred and seventy-fifth year of the Later Han starting from the first year of Chien-wu (A.D. 25).

INTRODUCTION

authorship is not known. It contains some seven hundred characters. The style resembles Liu Hsiang's *Pieh-lu* (Bibliographic Resumé). I surmise that it may be originally from Wang Chien's (452–489) *Ch'i-chih* and was later included in the *Han-chi*."[61]

Yao's conjecture is, however, rather ill-founded. For no author, neither Wang Chien nor even his predecessor Cheng Mo (who produced the *Chung-ching, Catalogue of Books in the Palace*, ca. A.D. 250),[62] who lived under a different dynasty, would have written such panegyrics to the Han rulers as appear in Preface A.

On the other hand, Yao's comment on the style of this document deserves credit. As discussed earlier, the impersonal and explanatory expressions in this document make it sound like an official bibliographical resumé. According to *Sui-shu* 34, the earliest then-surviving bibliographic record of the post-Han imperial libraries was the *Chung-ching* produced by Cheng Mo, an Attendant (*lang*) at the *Mi-shu-chien* of the Ts'ao-Wei regime.[63] In *Chin-shu* 44:2a, it is mentioned that Cheng Mo merely reedited some of the old documents (*chiu-wen* 舊文) treasured in the *Mi-shu-chien*. Since the *Mi-shu-chien* of the Ts'ao-Wei regime was an immediate successor to the bureau headed by Hsün Yüeh, these old documents appear to be none other than those collected and edited by Hsün Yüeh himself under the patronage of the Han emperor Hsien-ti, who, as previously noted, was markedly "interested in literature and talented in letters." Those sections in Preface A which contain the contemporary eulogies to the Han rulers no doubt belong to one of these old documents. It is likely that when these documents were reedited by Cheng Mo and others, some explanatory remarks were added to them which later by mistake merged into the text proper.

Since Hsün Yüeh was by that time in sole charge of the Han *Mi-shu-chien*, assisted only by a few copiers selected from the

---

[61] P. 220. This passage is followed by a quotation of Wang Ming-sheng's comment on the *Han-chi hsü*, which is also somewhat misleading.
[62] *Chin-shu* 44:2a; *Sui-shu* 32:4a.
[63] *Ibid.*

INTRODUCTION

emperor's bodyguards, he himself might actually have written this resumé of the *Han-chi*.⁶⁴ The attribution of Preface A to Hsün Yüeh thus accords with the general practice of the time, which restored the authorship of a document to the person who wrote it for the emperor or empress, who were its supposed authors.⁶⁵

This explanation may also solve the most difficult problem concerning the authenticity of Preface A., i.e., the difference in tone between this document and Preface B. For, in writing Preface B, Hsün Yüeh was addressing a memorial to the throne and accordingly exercising a higher degree of self-restraint. But when he wrote the resumé in Preface A, he was producing a memorandum vicariously for the Han court, and was therefore free to exalt his ideal with vigor and flourish.

For lack of more relevant external evidence, our critical appraisal of Preface A can be made only upon internal analysis. Generally speaking, the literary style of this document, with its frequent parallelisms, is highly characteristic of Hsün Yüeh's other writings in the *Han-chi*, and to a lesser extent in the *Shen-chien* as well. The views on the nature and function of historical writings propounded in Preface A are also in accord with Hsün Yüeh's ideas in Preface B and in *Shen-chien* 2:11b–12a, 13a, 15–16.⁶⁶

According to Preface A, Hsün Yüeh was appointed Custodian of the Secret Archives sometime before A.D. 198 and was commissioned by Emperor Hsien to compile the *Han-chi* in A.D. 198. He finished the work in the fifth year of Chien-an (A.D. 200), some five years before he submitted the *Shen-chien* to the throne. (Wang Ying-lin of the Sung dynasty accepted these dates in his encyclopedic work *Yü-hai*.)⁶⁷ The five inter-

⁶⁴ *Hou Han-shu* 62:14b; *Han-chi hsü*, 1b.

⁶⁵ For instance, see *Wen-hsüan* (*Ssu-pu ts'ung-k'an* ed.) 36:1–28. In the facsimile of a Sung manuscript (1131–1160?) of the *Han-chi* described in the *T'ieh-ch'in t'ung-chien-lou ts'ang-shu mu-lu* 9:1, this resumé had already been designated as *Tzu-hsü* 自序 (Author's Preface).

⁶⁶ For further discussion of Hsün Yüeh's literary style and his historiography, see *Hsün Yüeh*, ch. 5–6.

⁶⁷ *Yü-hai* (1687–1688 ed.) 47:17b–18a.

INTRODUCTION

vening years between 200 and 205 are important not only with regard to Hsün Yüeh's life and ideas, but also in terms of the manifold political and social changes that occurred during this early part of the Han-Wei transition period.

III. Selections from Hsün Yüeh's *Lun* (Discourses) in the *Han-chi*

The *Han-chi* and the *Shen-chien*, written in A.D. 198–200 and 200–205 respectively, record the reflections of Hsün Yüeh on history and on the issues of his time. Because the versions of the *Han-chi* that have been transmitted to us may contain passages that were originally part of the *Shen-chien*, as suggested by the discussion above, it is imperative to study the two works side by side so as to fully understand Hsün Yüeh's thought.[1]

An English translation of the entire *Han-chi* would fill several volumes, and the discourses (*lun*) alone would constitute a book nearly double the size of the present one. Since most of the discourses comment on events recorded in the *Han-chi*, they cannot be properly comprehended without extensive reference to the rest of the text. However, the selections presented here may be read as independent essays in themselves. They come from some of the longer *lun* in the *Han-chi* and deal with questions of history, cosmology, human nature, morality, and government policy. These excerpts complement the shorter and, in some cases, fragmentary passages on similar topics in the *Shen-chien*—providing an opportunity for comparison of the two works.

The introduction preceding the translation of each *lun* summarizes its content and describes its context in the *Han-chi*. Words left out of the original are inserted in brackets within the text, and in parentheses are added remarks that clarify the

---

[1] For a comprehensive study of Hsün Yüeh's life and thought, including a detailed discussion of his idea of history and the historiography of the *Han-chi*, see *Hsün Yüeh*.

INTRODUCTION

meaning of passages or briefly explain historical allusions that would otherwise require lengthy and cumbersome footnotes.

On Historical Situations (Selection one)

The first *lun* in the *Han-chi* draws a lesson from the history of Liu Pang, founder of the Han dynasty, who participated in the rebellion that vanquished the Ch'in dynasty in 206 B.C.[2] Emerging as the King of Han, he turned to the struggle with Hsiang Yü (the King of Ch'u), another rebel leader.[3] In 204 B.C., at the height of the battle between Hsiang and Liu, a Confucian scholar advised Liu to revive the six pre-Ch'in feudal states in order to gain additional allies.[4] Liu Pang's Taoist advisor, Chang Liang, opposed the proposal and as a result the king rejected it.[5]

Hsün Yüeh uses this example to discuss the complicated factors that affect the course of an event. He argues that times and conditions are ever-changing, and that no simple or fixed rule may be followed in all circumstances.

Han-chi 2 : 12b–14a

Hsün Yüeh says:

"The method of planning successful strategies consists of [the comprehension of] three important factors: first, *hsing* 形 (the general conditions); second, *shih* 勢 (the specific situation); third, *ch'ing* 情 (the conditions of men). *Hsing* means the overall favorable or unfavorable conditions; *shih* means that which is appropriate at the moment and makes a time propitious for

[2] Cf. *Han-shu* 1A–B; H. H. Dubs, *History of the Former Han Dynasty*, Vol. I, pp. 27–150.
[3] Cf. *Shih-chi* (*Po-na* edition), 7; Burton Watson, *Records of the Grand Historian of China*, I, pp. 37–74.
[4] The Confucian scholar is Li I-chi; cf. Watson, I, pp. 269–275. The advice is not recorded in his biography, but is mentioned in the biography of the Marquis of Liu (Chang Liang); see the following note.
[5] *Shih-chi* 55; Watson, I, pp. 134–151.

advancing or retreating; *ch'ing* means the mind and the intention [of men], which may or may not be appropriate [for the task]. The same strategy adopted in similar tasks may produce different results. Why? [This is because in each case any of] the three factors involved may be different.

"Previous [to the episode of 204 B.C.], Chang Erh and Ch'en Yü (two Confucianists who supported Ch'en She's uprising against the Ch'in in 209 B.C.) had advised Ch'en She to revive the six feudal states in order to establish more allies.[6] Now [the Confucian] Master Li advised the King of Han to do the same. Since the two sets of advice contained the same argument, why would one be good and one bad? The reason is that when Ch'en She staged the uprising (in 209 B.C.), all men under Heaven wished to destroy the Ch'in; but when the Ch'u (Hsiang Yü's group) and the Han [fought against one another] in the divided realm, all men under Heaven did not necessarily wish to destroy Hsiang [Yü]. Furthermore, before Hsiang Yü led the six [resurgent] states to attack and destroy the powerful Ch'in regime (in 206 B.C.), the *shih* [of Ch'en She] was incapable [of accomplishing this] (the general condition of the time was such that by his own effort Ch'en She would have been unable to defeat the strong army of the Ch'in). Therefore, Ch'en She's revival of the six former feudal states meant increasing the number of his allies and multiplying the opponents of the Ch'in. Besides, at that time Ch'en She did not control the whole realm [which was still ruled by the Ch'in]. What he did was to claim those [areas] which were not actually his and to give them [as fiefs] to others. This is to give unreal benevolence in order to beget real benefit.

"Now, for the King of Han to reestablish the six feudal states would have been to concede what he had possessed to the enemy

---

[6] Ch'en She's uprising in 209 B.C. sparked the anti-Ch'in revolution to which many ex-nobles and influential personages of the six former feudal states responded. These included Chang Erh and Ch'en Yü, two nobles from the Wei state. Ch'en She's forces were crushed by the Ch'in imperial army in 208 B.C., but the allied forces of the ex-nobles and other insurgents eventually overthrew Ch'in rule in 206 B.C. See Watson, I, pp. 19–33, 171–188.

INTRODUCTION

(in 204 B.C. the Ch'u and the Han had divided the realm between them, and the King of Han had to assign fiefs from his own domain to the ex-nobles of the six states, most of whom were allies of the Ch'u). This is to set up a titular [institution of authority] and to suffer real reverses from it. These two cases (the plan to revive the feudal states by Ch'en She and by Liu Pang) constitute what I call 'similar tasks in different *hsing*.'"

(Here Hsün Yüeh proceeds to discuss the "waiting strategy" used on two different occasions. In one case, Master Chuang of Pien, an ancient hero who was fond of tiger hunting, was advised to watch two tigers eating an ox until they engaged in a fight in which one was killed and the other was wounded; as a result, he procured both tigers without much effort on his part.[7] A similar "waiting strategy" was advocated by Sung I, the commander of the Ch'u army in 208 B.C. Sung I wanted to hold back the advance of the Ch'u insurgents and wait for the strong Ch'in army to exhaust itself in fighting the other insurgent groups. The strategy was severely criticized by Hsiang Yü, who killed Sung I in 207 B.C. and led the Ch'u group in an immediate attack on the Ch'in.)[8]

" ... The advice in these cases was similar. If it was given in the period of the Warring States (fifth to third centuries B.C.) when armed attacks from the neighboring states were not as desperate and sudden [as in later times], it would have been acceptable. These Warring States had been established for a long time; to them victory or defeat in a single battle was not a matter of life or death; their *shih* was incapable of quickly vanquishing a rival state (none of these states was strong enough or had so urgent a need to destroy another state in a single battle; and none was so weak as to be in danger of being

---

[7] The dates of the legendary tiger-hunter are uncertain. The story of his "waiting strategy" was recounted by Ch'en Chen, who advised King Hui of Ch'in (r. 337–311 B.C.) to wait for the two warring states of Han and Wei to exhaust themselves in mutual struggle before attacking them (*Shih-chi* 70:21); hence Hsün Yüeh's comment on the situation of the Warring States.

[8] Watson, I, pp. 44–46.

INTRODUCTION

eliminated in such a battle). They might advance for an attack and thereby make small gains, or else retreat to safety. Therefore they accumulated their strength and waited until they could take on the enemy at a deadly moment. This is due to their *shih* being as it was.

"Now (in 208 B.C.), the insurgents of the Chao (the group that was fighting the strong Ch'in army when the Ch'u leader advocated a 'waiting strategy') and of the Ch'u were in the *shih* (moment) of a life-or-death struggle with the Ch'in [imperial force]. Their chance for safety or danger changed in the time of a single breath. They either had to advance and succeed, or retreat and suffer the disaster [of being exterminated by the Ch'in]. These two instances (the 'waiting strategy' of the Warring States and that of the insurgents of 208 B.C.) constitute what I call 'similar tasks in different *shih*.'"

(Hsün Yüeh goes on to describe two battles fought by the Han forces in 205 B.C. and 204 B.C. respectively. In 206 B.C. Hsiang Yü had led a strong expedition to the Shantung area in the northeast. While he was still fighting there a year later, the King of Han made an attack from the northwest, penetrating deep into Ch'u territory in the Huai river valley and capturing Hsiang Yü's capital at P'eng-ch'eng. The Han troops were celebrating their victory when Hsiang Yü returned with a small army, attacked, and routed them. Many Han soldiers were pushed into the Sui River [a tributary of the Huai] and drowned.[9] Later the King of Han received valuable support from an able general, Han Hsin, who diverted Hsiang Yü's attention by attacking those feudal states allied with the Ch'u. In the second battle in 204 B.C., Han Hsin commanded a small army against the Chao state in the north, which was a powerful ally of the Ch'u. He sent his main forces across the Mien-man River [the river, in present Hopei Province, has long since disappeared] to form a battle line with its back to the river, but concealed a small group of cavalry to attack via a detour. When pushed to the waterfront, Han Hsin's soldiers fought a desperate

[9] Cf. Watson, I, p. 97.

INTRODUCTION

battle and withstood the Chao army, which was later surprised by the hidden cavalry and defeated near the Ch'in River [a tributary of the Mien-man].[10] In both cases, the Han army fought with its back to a river. In one, it suffered a disastrous defeat; in the other, it won a great victory.)

"In the battle against the Chao (in 204 B.C.), [the Han commander] Han Hsin stationed his forces with their backs to the Ch'in River but the Chao force still could not defeat him. In the disastrous [battle] at P'eng-ch'eng (in 205 B.C.), the King of Han fought with his back to the Sui River; his soldiers all fled into the river; and the Ch'u soldiers won a great victory. Why?

"The Chao soldiers went out from their city to meet the enemy in battle; they could advance when the opportunity was favorable and retreat [back to the city] when they knew [the battle] was going to be difficult. They were looking backward in their minds and had no determination to die [in the battle]. Han Hsin's forces were isolated on the bank of the river; the soldiers knew they must [win or] die, and had no other alternatives. This is why Han Hsin won.

"The King of Han [led his forces and captured the Ch'u capital P'eng-ch'eng by surprise. They were] deep in enemy territory and yet they drank wine in an arrogant festival [as a celebration of their easy victory]. The common soldiers were in a relaxed mood and had no firm determination to fight in [the more difficult] battle yet [to come]. The Ch'u, a great and mighty state, had just lost its capital city [while its leader, Hsiang Yü, was fighting elsewhere]. Hsiang Yü was returning to his state [set on recovering the city]; his soldiers were indignant, anxious to recover the defeated and the ruined, and ready to risk all in the coming battle. This is why the Han forces were defeated (in 205 B.C.).

"Besides, [in 204 B.C.] Han Hsin selected the best of his soldiers to defend [the battle line on the river bank] while the Chao sent out its hesitant soldiers to attack the Han; whereas

[10] Cf. Watson, I, pp. 214–217.

INTRODUCTION

[in 205 B.C.] Hsiang Yü selected his best soldiers for the attack while the Han sent its insolent and over-confident soldiers to respond. These [two battles] constitute what I call 'similar tasks of different *ch'ing*.'

"Therefore, it may be said [in conclusion] that what is expedient [under different circumstances, *ch'üan* 權] cannot be anticipated and what is ever-changing cannot be charted ahead of time. To change according to the times and to [always] be responsive to the modification of events is the key to strategic planning."

On the Correspondence Between Nature and Man (Selection Two)

An eclipse of the sun was recorded on the *chi-ch'ou* day of the first month of the seventh year (March 4, 181 B.C.) of the Empress Lü's reign. The eclipse was considered a bad omen for the empress.[11] Appended to the entry in the *Han-chi* about this event is an excerpt taken from the "Treatise on the Five Elements (*wu-hsing*)" of the *Han-shu*. It is followed by Hsün Yüeh's *lun*, in which he reaffirms the correspondence between the cosmos and man, and criticizes those who followed Wang Ch'ung (A.D. 27–91) in denying any such relationship. Hsün Yüeh argues that the correlation is too subtle for men to comprehend. He then shifts his discussion to the nature and conditions of man.

Han-chi 6: 4a–6

Hsün Yüeh says:

"All changes and abnormalities of the Three Luminaries (the sun, the moon, and the stars) and of the spirits and ethers (*ching-ch'i* 精氣) reflect the essence of *yin* and *yang*. These originate from the earth and reveal themselves in Heaven. When maladministration occurs here [on earth], change

[11] Cf. Dubs, I, p. 199.

INTRODUCTION

appears there [in Heaven]. This is like a shadow simulating the shape [of the object which casts the shadow], or an echo responding to a sound. Therefore, an enlightened ruler who sees [such an omen] and understands [its meaning] will reprimand and rectify himself; he will ponder his mistakes and sincerely apologize for his faults; then calamity will disappear and good fortune will be generated. This is a natural correspondence.

"However, the Ode reads: 'The doings of High Heaven have neither sound nor smell.'[12] Their details are difficult to know, are they not? Bad or good omens are sometimes correspondent [to men's deeds] and sometimes not. According to the interpretation of bad omens in the "Great Plan" (*Hung-fan*), [disastrous flood and drought would befall the bad rulers];[13] yet [the Sage-rulers] Yao and T'ang suffered the disasters of flood and drought.[14] Tradition [teaches that man's efforts at reform should] dispel disasters and restore good omens; yet after King Hsüan of the Chou Dynasty [had restored the virtue of his dynasty], clouds and drought [persisted in their visitations].[15] All this does not coincide with our [sense of the correspondence between] virtuous conduct [and fortunate omens].[16] According to the tradition of the *I-ching* (*Book of*

---

[12] James Legge, *The Chinese Classics*, IV, p. 431.

[13] The "Great Plan", in *Shu-ching* (*Book of History*, Legge, III, pp. 320–344), was traditionally considered to be the counsel on statecraft given to the early Chou ruler by a surviving Shang noble. Although in fact much later in date, it contains what was supposed to be the earliest tradition of the "Five Elements" theory. The elaboration of this theory by scholars of the Former Han dynasty is recorded in *Han-shu* 27A-B-C.

[14] The great flood during King Yao's reign (traditional dates, 2145–2043 B.C.); cf. Legge, III, pp. 24–25. According to the *Annals of the Bamboo Books*, during King T'ang's reign the Shang dynasty suffered a great drought for six years (traditional 1555–1550 B.C.) until he prayed in the Mulberry Forest; Legge, III, Prolegomena, p. 129.

[15] According to the *Annals of the Bamboo Books*, a drought lasted for some five years (traditional 830–826 B.C.) prior to King Hsüan's restoration. Under King Hsüan there was also a drought, in the twenty-fifth year of his reign (802 B.C.); Legge, III, Prolegomena, pp. 154–156.

[16] These were mentioned by Wang Ch'ung in *Lun-heng* (*Han-Wei ts'ung-shu*

INTRODUCTION

Changes), those who accumulate virtue will have good fortune;[17] yet Yen [Hui] and Jan [Keng] suffered the calamities of premature death and chronic disease respectively.[18] The effects of good or evil [doings] and the factors affecting events and matters change in myriad ways; these cannot be reduced to a single [principle]. Therefore, those who see and hear (those who depend upon their common sense alone to find proof for the correspondence between Heaven and man) are confused.

"If one follows the laws (*shu* 數, numbers, numerological rules) of nature and investigates the reasons (*li* 理, the underlying principle) for man's nature and fate, studies these in the canonical classics and examines them [in the events of] the past and present, encompasses the three conditioning factors (*shih*) and penetrates to the subtlest essence [of things], holds the two extremes and controls the mean, traces the changes in the three [primal spheres: Heaven, Earth, and Man] and in the five [elements: Metal, Wood, Water, Fire, and Earth], and places these in [their correct and] complicated combinations and sequences, then one may dimly perceive the [true meaning of] an omen.

"Generally speaking, the nature of things is such that some [events] naturally occur of themselves and some of them wait for man's effort in order to occur, while some of them will not occur in spite of man's life-long efforts. These are called the three conditioning factors (*shih*). The three conditioning factors apply to all things. [We may infer the nature of their operation] from the small to the great. [Let us] use a proximate example, our body, and look at [the question of] disease. Some diseases need no treatment to be cured; some are treated and cured; some are not treated and, therefore, are not cured; some

---

edition, First Series), 5:15–17, 15:11–12b, and 17:17, to refute the theory of correspondence between nature and men.

[17] *I-ching* 1:26a; tr. by Legge, *The Sacred Books of the East*, edited by F. Max Muller, Vol. XVI, p. 419.

[18] Yen Hui, considered by Confucius to be his best disciple, suffered a premature death; Jan Keng (Po-niu), another disciple of Confucius, suffered from a chronic disease. See Legge, I, pp. 185 and 188.

are not cured despite life-long treatments. In ancient times, the prince heir-apparent of Kuo was dying; Pien Ch'ueh treated him and revived his life. Ch'ueh said: 'I am not able to bring the dead to life; I am only able to keep alive those who can be made to live.'[19] However, if the prince had not met Ch'ueh, he would not have remained alive. As to those deadly diseases which affect the vital organs (*kao-huang*) of men, they cannot be healed even by the best physician.[20] Therefore, Confucius said: 'Death and life have their determined appointments.'[21] He also said that some would not die a natural death[22] and that some had fortunately escaped death.[23] The statement, 'Death and life have their determined appointments' refers to the principles [regulating life and death]; 'some do not die a natural death' refers to those who die [at a moment or in a manner] in which they should not have died; 'some fortunately escape death' refers to those who should have died and yet had not. All these derive from the cause of the three conditioning factors (*shih*) in [man's] nature and fate.

"Reasoning from this to the effect of education, [one may see that the factors are] similar. Why is it so? Some men need no education to fulfill [their potential]; some wait for education to fulfill [their potential] and cannot do so without education; some cannot fulfill [their potential] in spite of life-long education. Therefore, the highly intelligent and the lowly foolish will not change.[24] But those men in the middle can be either elevated or lowered.

"Reasoning from this to the Way of Heaven, [one may see that the factors] are similar. And the correspondence of the bad

---

[19] See *Shih-chi* 105:3-6.
[20] Cf. *Shih-chi* 105:6b-7a.
[21] This comes from the saying of Tzu Hsia (Pu Shang), one of Confucius' disciples, reporting what he had probably heard from Confucius; Legge, I, p. 253.
[22] Cf. Legge, I, p. 241.
[23] Cf. Legge, I, p. 190.
[24] This may also refer to Confucius' sayings in the *Analects*, Legge, I, pp. 313-314 and 318.

or good omens [to man's affairs] will not be mistaken. The flood and drought in the times of Yao and T'ang belong to the natural laws of Heaven; the omens mentioned in the "Great Plan" relate to the affairs of man; the good rain in the time of the Duke Hsi of Lu is the effect of the probable (something which can be affected by man's effort);[25] the drought in the time of King Hsüan of Chou is a situation that would be difficult for men to change; the calamities of Yen [Hui] and Jan [Keng] originate from the nature and fate of man (i.e., they are unalterable physiological conditions). All this may be compared to the movement of the celestial realm which encompasses the cycling of the sun, a great movement that pushes and propels [everything in it]. Although this may be called "chance" (*yü* 遇), good and bad fortune are, nonetheless, involved.

"Now, there are those who see something which cannot be changed [by man] and say that man's efforts cannot change anything; there are those who see something which can be changed [by man] and say that nothing is destined by Heaven; there are those who see that Heaven and man are distant from each other and say that man's efforts have no effect [on the other sphere]; and there are those who are aware of the transcendence of spirits and ethers (*shen-ch'i* 神氣) and thus [hold the view that] men who undertake similar tasks will produce similar results. Each of these scholars adheres to one point and does not examine [the whole] from beginning to end.

"The *I-ching* states: 'There is the Way of Heaven; there is the Way of Earth; and there is the Way of Man.'[26] This refers to the distinctions [of the three spheres]. It (the *I-ching*) synchronizes the three [spheres] by the two [principles, *yin* and *yang*]. This refers to the similarities [of the three spheres]. The Way

---

[25] The *Spring-and-Autumn Annals* record that it did not rain during the first and fourth months of the third year (656 B.C.) of the reign of Duke Hsi of the Lu state, but that it rained during the sixth month of that year; Legge, v, p. 137. According to the Kung-yang tradition, the rain came in response to the Duke's effort at self-rectification; *Ch'un-ch'iu Kung-yang chu-shu* (I-wen yin-shu-kuan reprint of the 1815 edition), 10:10–11.

[26] *I-ching* 8:22a; Legge, *The Yi-king*, p. 402.

INTRODUCTION

of Heaven and the Way of Man have their similarities and their differences. If one bases [his investigation] on the differences and seeks the similarities, [his conclusions] will be satisfactory. If one bases his investigation on the similarities and seeks the differences, [his conclusions] will be unclear. (Here Hsün Yüeh tends to disfavor thinking by way of deduction from the abstract to the particular.) Thus, Confucius said: 'There is love of knowing without the love of learning—the beclouding here leads to dissipation of mind.'[27]

"The later vulgar [generation of men] observes the complicated and disorderly [nature of situations] and the confusion and discord in the affairs [of man] and in the changes [of the cosmos]. Different ideas spring forth in disarray; they lose their underpinnings. Opinions that are erroneous, scattered, and opposed to the Way are generated; ideas that malign the spirit and denigrate the Sage are produced. However, although the highly intelligent and the lowly foolish cannot be changed, there are many [men] who can be changed through education; although the greatest law [of nature] does not change, there are many [things] that can be changed by man's efforts. Furthermore, there are some diseased persons who have been treated but not yet been cured, or who have been cured but have not yet healed, or who have healed but have not yet fully recovered. [With regard to] education, there are [cases] in which [people] have been taught but have not yet been moved to action, or who have been moved to action but have not yet accomplished their goal, or who have accomplished their goal and then have been corrupted. The corresponding spirits may have moved but not yet responded [to men's affairs], or may have responded but not yet completed [their response], or may have completed [their response] but then have changed (moved in other directions). [The progress of this movement] may be slow or fast, [its effects] may be shallow or profound, and [the process may result in] complicated mutations. The elements [involved] are extremely complex and diverse, and [the course of their interac-

[27] Legge, I, p. 322.

INTRODUCTION

tion is] difficult to ascribe [to any one single rule]. The principles (*li*) of Heaven, Earth, Man, and the matters [of the world] are all like this.

"The laws (*shu*) of the three conditioning factors (*shih*) are too deep for man to comprehend. Therefore, a superior man should simply exert his mental strength to the utmost, and then follow what is destined by Heaven (*ming* 命). The *I-ching* states: 'Seek the basic principles (*li*) to the end and develop man's nature (*hsing*) to the utmost until you reach [what is determined by] fate (*ming*).'[28] Is this not what it means?"

On Land Policy (Selection Three)

In the sixth lunar month of the thirteenth year of Emperor Wen's reign (167 B.C.), a decree was issued abolishing the land tax on cultivated fields.[29] The *lun* on land policy appended by Hsün Yüeh to the notation of this event in his chronicle was written when China was on the eve of the Age of Disunity. It represents the opinion of a prominent scholar-official on an issue that was a major cause of the Han dynasty's decline. Hsün Yüeh's views on the restriction of land ownership and the "well-field" system have additional significance in light of the land policies followed by subsequent regimes.[30]

---

[28] *I-ching* 9:3a; Legge's translation in *The Yi-king*, p. 422, is too definitive.

[29] See Dubs, I, p. 255. The land tax was reintroduced in 156 B.C. at the rate of one-thirtieth of the product; Dubs, I, p. 311. The exemption and later the low tax rate did very little to relieve the plight of the poor peasants, for many of them were tenants of the big landlords, to whom they paid rent at the rate of about one half of their crop. The measures greatly benefitted the landlords who were responsible for payment of the land tax to the government. In addition, many wealthy merchants were induced to become landlords by reinvesting their money in land, thus aggravating the crisis of land encroachment. The crisis came up for court discussion as early as 163 B.C.; Dubs, I, pp. 261–262. For other details on the problem of landlordism in Han times, see Nancy Lee Swann, *Food and Money in Ancient China* (Princeton University Press, 1950), pp. 148–216; L. S. Yang, *Studies in Chinese Institutional History* (Harvard University Press, 1963), pp. 131–140.

[30] For further discussions, see *Hsün Yüeh*, Chapter Two, Section 1, and Chapter Six, Section 3.

INTRODUCTION

This discourse also demonstrates how Hsün Yüeh's concept of time (*shih*) affected his political outlook. He attempted to strike a compromise between idealism and realism in terms of what government should and could do at a particular time in a particular situation. "Justice and equality" should be the moral aims of government, he contended. But he realized that in difficult historical periods, it is impractical or impossible for the courts to establish just and egalitarian laws dealing with land ownership. Nevertheless, he maintained that the ruler and his court must keep the higher goals of justice and equality firmly in mind, doing whatever they can even if it amounts to no more than a moral gesture or symbolic action. In this way, Hsün Yüeh reasoned, governmental policy will be set on the proper course so that in more propitious times concrete steps will be taken towards the fulfillment of these goals. "Holding fast to the moral ideal while yielding in all practical matters" is the attitude that comes through in most of Hsün Yüeh's political comments in the *Han-chi* and *Shen-chien*, which were written during very chaotic and trying times.[31]

Han-chi 8 : 3–4a

Hsün Yüeh says:

"In ancient times, taxes were levied by the tithe.[32] This is considered the just mean for all [things] under Heaven. Now, the Han people were sometimes taxed at the rate of one-hundredth [of the product].[33] This was a very light [tax]. Yet the powerful,

---

[31] This is one of the major themes expounded in my first book on Hsün Yüeh.

[32] According to Mencius, the Hsia, Shang, and Chou dynasties all taxed their subjects at the rate of about one-tenth of the product (calculated by the size of, or the labor contribution to, the lord's field, *kung-t'ien*); Legge, II, pp. 240–241.

[33] At the beginning, the land tax in Han times was at the rate of about one-fifteenth of the product. In 168 B.C. the rate was reduced to one-thirtieth of the product. From 167 to 157 B.C. the land tax was abolished altogether. For the remaining years of the Han dynasty, it was at the rate of one-thirtieth of the product; see note 29. However, special reductions or exemptions were

INTRODUCTION

the strong, and the rich held land in excessive quantities and exacted rent [from their tenants] at the rate of more than one-half [of the product].[34] The government received taxes at the rate of one-hundredth and the people received rent at the rate of more than one-half. The benevolence of the government exceeded that of the three ancient dynasties (Hsia, Shang, and Chou); yet the powerful magnates were more cruel than the overthrown Ch'in rulers in their harsh treatment [of tenants]. Consequently, the benevolence from above did not reach [those below]; the power and profit [of the government] were abrogated and divided by the strong magnates. Now [Emperor Wen] did not rectify [the situation] at its root, but concentrated on exemption from and abolition of taxes. This only fostered the interest of the powerful magnates.

"According to the teaching of the *Spring-and-Autumn Annals*, the feudal lords ought not to have absolute power over their fiefs, and the grandees ought not to have absolute rights to their landholdings.[35] But [as it happened], the magnates sometimes held land amounting to hundreds and thousands of *ch'ing* (one *ch'ing* equals one hundred *mou*, or approximately thirteen English acres), and they became wealthier than the princes or marquises. This is nothing less than self-styled

---

granted to certain persons (the aged, the disabled, widowed, etc.), in certain regions (those affected by famine or other disasters), or on certain occasions (the enthronement of a new emperor, etc.). This is probably what Hsün Yüeh meant by the low tax rate of one-hundredth.

[34] See note 29.

[35] According to the Confucian ideal of political unity, only the sovereign ruler wields ultimate power, authority, and rights over the realm. Thus the Ode reads: "Under the wide Heavens, all is the King's land. Within the sea-boundaries of the land, all are the King's subjects." See Legge, IV, p. 361; also *ibid.*, V, pp. 615–616. From this ideal, the Kung-yang and Ku-liang Schools of Confucianism derived many of their criteria of praise-and-blame of the feudal nobility in the *Spring-and-Autumn Annals*. For instance, see *Ch'un-ch'iu Kung-yang chu-shu* 10:1b–2, 22:8a, and *Ch'un-ch iu Ku-liang chu-shu* (I-wen yin-shu-kuan reprint of the 1815 edition), 3:2b, criticizing those feudal lords who assigned sub-fiefs by themselves; also *Ch'un-chiu Kung-yang chu-shu* 24:20b–21a and 25:4b, on restricting the power or rights of the grandees.

feudalism. They bought and sold land at will, demonstrating fully their self-presumed rights to their property.

"During the reign of Emperor Wu (140–87 B.C.), Tung Chung-shu proposed limitations on private land ownership. Around the time of Emperor Ai (r. 6 B.C. to 1 A.D.), individuals were restricted from holding land in excess of thirty *ch'ing*. Although this regulation was issued, it was not enforced.[36] But even [the limit of] thirty *ch'ing* was not egalitarian.

"Generally, the "well-field" system (*ching-t'ien*)[37] is suitable in a time of overpopulation. [In a time] when there is plenty of land but very few people, the system may be discarded. Yet if this system is abandoned in [a time of] sparse [population] and introduced in [a time of] overpopulation, [then in the latter situation] the richly [yielding] fields would all be in the hands of the powerful magnates; and if this condition is abruptly altered [the magnates] will have resentment in their hearts, resulting in confusion and disorder. It will then be difficult to enforce the [egalitarian system].

"Regarding the above, [it may be noted that] when Emperor Kao-tsu had first pacified the realm or after Emperor Kuang-wu (r. A.D. 25–57) had restored the dynasty—the population being sparse—it would have been easy to establish an [egalitarian] system. Today, even if it is not possible to put into effect a full-fledged "well-field" system, a law limiting per capita landholding should be passed. People should be allowed to cultivate but not to buy or sell land. This would help the weak and prevent outright encroachment by the strong. Furthermore, it would lay the foundation for [a better] system to evolve. Is this not the proper [approach]? Although ancient and modern times are different and modifications [of the

---

[36] See note 29.

[37] The "well-field" system represented the Confucian ideal of an egalitarian land system which was supposed to have been realized during the Golden Age. The system was mentioned by Mencius (Legge, II, p. 245) and described in various ways in many other Confucian works. See Hsü Chung-shu, "The Well-field System in Shang and Chou," in Etu Zen Sun ed., *Chinese Social History* (Washington: American Council of Learned Societies, 1956), pp. 3–17.

system must be made] according to the conditions of the times, the primary aim of government should remain the same."

On Education and Law (Selection Four)

This is the second part of a long discourse (23:9b–11) following Hsün Yüeh's recording of the death of Emperor Yuan in 33 B.C. and the banishment and subsequent death of his favorite eunuch, Shih Hsien. The first part of the *lun* summarizes and evaluates the achievements of Emperors Kao-tsu (r. 206–195 B.C.), Wen (r. 179–157 B.C.), Wu (r. 140–87 B.C.), and Hsüan (r. 73–49 B.C.). Hsün Yüeh observes that Emperor Kao-tsu had an excellent personality, but his dynasty had just been founded and effective institutions were yet to be established; Emperor Wen was moderate and benevolent and attained peace during his reign, but the good traditions were yet to be revived and universal peace was yet to prevail; Emperor Wu fostered scholarship, achieved greater military power, initiated many reforms, and founded new institutions, but his accomplishments lacked substance (i.e., moral worth) and provoked many evils and much disorder and suffering; and Emperor Hsüan was a competent ruler and a good administrator, but he did not follow the Confucian teachings and, unfortunately, in the end favored the evil eunuch Shih Hsien.[38]

At this point Hsün Yüeh notes that Emperor Hsüan had once been advised by his heir-apparent, the future Emperor Yuan, to favor Confucianism. He did not heed this counsel, and predicted that it would be this very heir-apparent who would confound Han rule.[39] Here begins the second part of the discourse, which discusses the priority that ought to be given to education and law in different times and under different circumstances.

[38] For brief descriptions of the reigns of these emperors, see Dubs, I, pp. 2–25, 215–219; II, pp. 7–25, 180–198.

[39] Cf. Dubs, II, pp. 299–301; also pp. 341–353 (the victory of Han Confucianism).

INTRODUCTION

Han-chi 23:10b–11

[Hsün Yüeh's discourse reads]:

"..: Of those who debated [the question of effective] governmental rule, some said [the government should achieve moral] influence through education; some said there should be control through law; some said that education should precede statutory punishment [while] some said that statutory punishment should precede education; some said that moral education must be extensive [while] some said that it should be elementary; [some said that legal control should be detailed while] some said it should be simple; some said that statutory punishment should be light [while] some said it should be heavy. They all adduce [their doctrines] from one aspect of administration and do not comprehend the essence of administration from the beginning to the end; nor do they comprehend the great moral virtue of the Sage.

"The Way of the Sage must take its model from Heaven and Earth and be regulated by the Five Elements so as to comprehend and transcend change. It (the Way) is versatile and not dogmatic. The coexistence of moral [influence] and [legal] control is [ordained] in the permanent Ways of Heaven and Earth. The Way of the ancient [Sage-]rulers, however, exalts educational influence and denigrates legal control, preferring civil virtue to military merit—this is moral. Yet priority is sometimes given to moral education and sometimes to legal control—this is due to [the nature of] different situations (*yü* 遇). To avert disorder and to suppress the strong [and unruly], priority must be given to legal control; to assist the weak and to foster a new [way of] life [in the state or culture], priority must be given to educational influence. In a stable and peaceful realm, law and education are both used. In a time of great turmoil, there will be no education; in an era of great peace, there will be no legal control. That there is no education amidst great turmoil is due to the condition (*shih* 勢) [being such that education] is unavailable; that there is no legal control in [a period of great] peace is due to the time [being such that legal control] is no longer used.

"Education, at the early stage, must be easy. Legal control, at the beginning, must be simple. Their progress must be gradual. When the influence of education has reached a high point and there is no one who has not been elevated [to a state of good conduct], you may then charge it (education) with achieving perfection. When legal control has been firmly established and there is no one who has not been [warned] to avoid offending it, you may then demand that it (legal control) be more elaborate. [To charge moral education with the achievement of perfection] when it is not yet possible [for people] to become perfected is called excessive education. [To demand more elaborate] legal control when it is not yet possible to make it more elaborate is called cruel punitive [control]. Excessive education harms the effectiveness [of education]; cruel punitive [control] harms the people. A superior man does not follow either of these courses.

"To devise an educational system [that is, on such a complex level] that it alienates the people from it, without considering their inability to attain [the high standard demanded], is to push people into evil-doing. Therefore, it harms the effectiveness [of education]. To formulate laws [that are so extensive] as to make it impossible for people to avoid offenses, without considering [the fact that this makes] their condition unbearable, is to trap people into crime. Therefore, this harms the people. Only when all people have been elevated to [a life of] good conduct and it becomes possible to exhort them to be good in the most minute particulars can you then charge education with the attainment of perfection. Only when all people have been [warned] to avoid crime and it becomes possible to prohibit the slightest offense, can you then tighten and elaborate legal control.

"Confucius said: 'If he cannot govern with solemnity, the people will not respect him; when he governs with solemnity, yet if he tries to move the people contrary to the rule of propriety, excellence is not achieved.'[40] This refers to the use

---

[40] Legge, I, p. 303.

INTRODUCTION

of both the rule of propriety (*li* 禮, an important aspect of Confucian education) and punitive law. He also said: 'I can really do nothing with him.'[41] This refers to the impracticability of education [in certain cases]. He further said: 'They would be able to transform the violently bad and dispense with capital punishment.'[42] This refers to the dispensability of punitive law. The *Chou-li* reads: 'To govern a new-founded state, one must employ light statutory ordinances.'[43] This refers to the simplicity [of the judicial program] at the initial stage. The *Spring-and-Autumn* classic censured the slightest crime.[44] This refers to the ultimate elaboration [of judicial control]. Confucius said: 'When [a man], after performing [all the essential duties which are required of him simply as a man], still has [time and energy], he should employ them in polite studies.'[45] This refers to the simple beginning [in education]. He also said: 'The business of laying on the colors [as in decorative painting] follows [the preparation of] the plain ground.'[46] This refers to the perfection of the raw material in the final stage [of education].

"It is generally [the case that] when one comprehends the principles of Heaven and Man and understands the laws of change, one reaches the Way. The Sage emulates Heaven; the worthy learn the rules of Earth. By studying the Way of Heaven and making reference to the canons and the classics, one makes use of them in the proper manner."

[41] Legge, I, p. 224.
[42] Legge, I, p. 267.
[43] *Chou-li chu-shu* (I-wen yin-shu-kuan reprint of the 1815 edition), 34:13a.
[44] For a traditional interpretation of the *Spring-and-Autumn Annals*, see Legge, v, Prolegomena, pp. 1–16.
[45] Legge, I, p. 140.
[46] It means that learning, like painting, begins with laying out the raw background material and ends with finishing the subtlest details in the picture; see Legge, I, p. 157.

THE *SHEN-CHIEN*
(EXTENDED REFLECTIONS)

Shen-chien 1
Essence of Government
(Cheng-t'i)

1.1 (1a4) Humanity (*jen*) and Righteousness (*i*) are the foundation of the Way. The Five Classics are its warp, and all other writings its woof.¹ It manifests in poetic recitation, singing, stringed music, and dance.² The reflection of history (*chien*) is already clear.³ [Its meaning], however, needs to be

¹ The Five Classics (*ching*) consist of the *Book of Poetry* (*Shih*), the *Book of History* (*Shu*), the *Book of Changes* (*I*), the *Book of Rites* (*Li*), and the *Spring-and-Autumn Annals* (*Ch'un-ch'iu*). These were established as canonical works, *ching*, of the official Confucian orthodoxy under the Han dynasty. *Ching* in this sense means the "permanent" canons. *Ching* also has the broader meaning of "warp," which, together with the *wei*, or "woof," are said to constitute the cardinal threads of all things—whether a state, a society, or the cosmos. *Wei*, like *ching*, also has a more specific meaning. The *wei* were a set of apocryphal-prognostic works that emerged during the Han dynasty, allegedly representing the esoteric teachings of Confucius. This claim was repudiated by many important Confucians of the Later Han such as Hsün Shuang (A.D. 128–190), Hsün Yüeh's uncle, and Hsün Yüeh himself (see *SC* 3:15 below). Although Hsün Yüeh did not deny the value of the *wei* writings, he would not accept them as the cryptic words of Confucius and the only supplement to the canonical Classics. In the passage presented here, he upheld the Five Classics as the "warp" of the Way and all other books as its "woof." A similar attitude of compromise can be found in many discourses in the *Shen-chien*.

² For the importance of song and dance in ancient China, see Marcel Granet, *Festival and Songs of Ancient China*, tr. by E. D. Edwards (London, 1932).

³ 鑑, bronze mirror, often decorated with cryptic signs and joyous inscriptions. Its root character 監 signifies a person looking into a tray of water, hence to oversee, observe, and reflect. The variant 覽 means to see, to read. Thus the term conveys the notion of reflection, reflective thinking, learning by example and from the lessons of history. The founder of the Chou dynasty was said to have been deeply affected by the tragic example of the fallen Shang dynasty, and frequently admonished his people to be mindful of the lesson of the Shang, *Yin-chien* (lit. Mirror of the Yin—Yin being another name for the

extended and restated in later times.[4] Therefore the ancient Sage-kings devoted themselves to extending and stressing Humanity and Righteousness. Faithfully following this through time everlasting is what I call *Shen-chien* (Extended Reflections).[5]

1.2 (1a7) The sacred Han [dynasty] receives the Mandate of Heaven.[6] Verily it reveres and assists in this timely work,[7] resulting in merits encompassing the universe.[8] There are valiant ministers who bring order [out of disorder];[9] wastes and ruins are recovered in time, according to the grand

---

Shang dynasty). Such reflective wisdom (historical knowledge in a peculiar sense) had been construed as one of the factors enabling the Chou to receive the Mandate of Heaven, a concept vigorously espoused by the founders of the Chou to legitimize their conquest of the Shang. Hence the loss of *chien*, i.e., ignorance of historical lessons, would doom a dynasty.

[4] *Shen*, to notify, to order, or to extend and to repeat. Here the appearance of the character *fu* (again) clearly denotes the act of reiterating.

[5] For the translation of *Shen-chien* as "Extended Reflections," see Introduction.

[6] The paragraph beginning with "*Sheng-Han*" consists of fourteen sentences in an extremely archaic style. Most of them, as will be seen in the following notes, are quotations from the *Shu-ching* (the *Book of History*), while others are not direct quotations but allusions or new combinations consisting of a number of archaic characters from the same work. "*Sheng-Han t'ung-t'ien*" may be rendered either as "the sacred Han commanded [all under] Heaven" or "the sacred Han acquired her lineage from Heaven" (*t'ung-t'ien* or *t'ien-t'ung*, the heavenly lineage). The paragraph may be compared to the passage in *Shih-chi* 8:39b, 漢興 ... 得天統矣, "The Han rises ... [and] acquires her lineage from Heaven." Cf. Watson, I, pp. 118-119.

[7] 惟宗時亮, *wei*, to think, to think only, only; *tsung*, to honor, to worship one's ancestors, ancestral lines; *shih*, time, this time, this; *liang*, to aid, brilliant, to display brilliantly; Legge, III, pp. 649, 667-8, 667, 684. The whole sentence is a compressed quotation from the "Canon of Shun," Legge, III, p. 50 and *pass*.

[8] *Ko*, to reach, to correct, most excellent; Legge, III, p. 687. This sentence is a compressed quotation from the "Canon of Yao," Legge, III, pp. 15-17.

[9] *Hu-ch'en*, lit., "Tiger Ministers"; *luan-cheng*, lit. "Disorderly Administration." However, *luan ch'en* 亂臣 (lit., disorderly ministers) also alludes to the "ten capable governing ministers" who helped King Wu found the Chou dynasty (Legge, III, p. 292). Thus the sentence has the literal meaning of "There are Tiger Ministers confounding the administration," and also makes a diametrically opposed allusion to "the valiant ministers capable of orderly

plan;[10] and these (the institutions of the sacred Han dynasty) reflect the model and the canons of the Three Dynasties.[11] If the ruler duly increases his moral strength,[12] his meritorious deeds will excel even more as time goes on.[13] The Heavenly Way is here;[14] the emperor need only exert himself toward it.[15] Grace will visit the high and the low,[16] and peace will prevail in the myriad states. All these spring forth [from the Way]. Oh, how far-reaching![17]

---

administration." Here Hsün Yüeh seems to be intentionally ambivalent. The "Tiger Minister" probably refers to Ts'ao Ts'ao (155–220), the strong man behind the *Chien-an* regime (196–220) of the Later Han. Ostensibly, Ts'ao was working for the restoration of the Later Han dynasty; but actually, he had established a *de facto* regime of his own which became the Wei dynasty that replaced the Later Han in A.D. 221. When Hsün Yüeh wrote the *Shen-chien* (sometime between A.D. 196 and 205), he was hoping that Ts'ao might still be persuaded to change his attitude from that of a "Tiger Minister" subverting the Han regime to that of a "valiant minister" supporting a genuine restoration of the dynasty. The next sentence in the passage evinces a similar ambivalence between its literal meaning and its historical allusion. For further discussion of this point, see *Hsün Yüeh*, pp. 130–132.

[10] See the preceding note. *Huang* 荒, waste, vast; *chi* 圮, ruin; *yen* 湮 or 堙, to obstruct, to destroy. This is an allusion to a devastating flood under the rule of King Shun (r. 2255–2205 B.C. by tradition); Legge, III, pp. 25, 323. The disaster led to the achievements of the Sage-king Yü of the legendary Hsia dynasty, who reclaimed the land that had been ruined by the flood.

[11] Here the key word *chien* appears once more with the meaning of "learning from the past." The Three Dynasties were the Hsia, Shang, and Chou. The character *tien* (precedents, classics) may refer either to the institutions of these ancient dynasties, which were looked to as models, or to the Classics, which were supposed to contain the records of these institutions.

[12] From the "Counsels of Kao-yao," Legge, III, p. 68.

[13] *Yu-shang* connotes something that is at a peak of excellence, but which still may be improved or superceded.

[14] This is a compressed quotation from the "Counsels of the Great Yü," Legge, III, p. 61: "The determinate appointment of Heaven rests on your person."

[15] An allusion to the "Counsels of the Great Yü" and the "Counsels of Kao-yao," Legge, III, pp. 59, 74.

[16] The term *she-chiang* is generally employed in sacrificial hymns to praise the virtues and blessings of the ancestor of a dynasty; Legge, IV, pp. 428, 485.

[17] The paragraph ending here consists of quotations, allusions, and in-

1.3 (1b3) The Way which informs Heaven is called *yin* and and *yang*;[18] the Way which informs Earth is called hard and soft;[19] the Way which informs Human Beings is called Human-heartedness (*jen*) and Righteousness (*i*).[20] *Yin* and *yang* determine the spiritual essence [of Heaven]; hardness and softness distinguish the myriad shapes [on Earth]; Humanity and Righteousness constitute the meritorious deeds [of man].[21] This is the Way.

Consequently, the great principles of government are law and education. Education is the transforming force of *yang*; law is the embodiment of *yin*. Kindhearted is he who cherishes these, righteous is he who accords with these, decorous is he who practices these, faithful is he who abides in these, and wise is he who knows these.[22] These (education and law) are manifested in the feelings of like and dislike, determined by the feelings of cheerfulness or anger, and sustained by the feelings of joy and sorrow.

---

dividual characters taken from the *Shu-ching*. Most of them are praises to a group of Sage-kings of remote antiquity beginning with Yao and Shun. Apparently Hsün Yüeh thought that including such a compendium of eulogies would serve to link the great Sage-kings of antiquity with the early emperor or emperors of the Han. The eulogies were thus a compact way of presenting what he called the *chien*, the "historical lessons to be learned from the ancient Sage-kings" as embodied in the "Historical Documents." For further discussion, see *Hsün Yüeh*, pp. 129–132.

[18] The implied subject in this and the next two sentences is unclear. It could be the emperor, as referred to in the last sentence of the preceding paragraph; thus the sentence may be construed to read, "The emperor established the Way of Heaven and called it *yin* and *yang*," etc. However, this suggests a strongly anthropocentric view of the *tao* (Way), a view which Hsün Yüeh repudiated.

[19] The *yang* principle of Heaven is analogous to the quality of *kang* (strength, hardness) of the Earth and the quality of *i* (moral justice, righteousness) in the human world; *yin* is analogous to the *yu* (soft, weak) fluid of the Earth and *jen* (the virtue of human-heartedness) in human beings.

[20] From the *I-ching*, Legge, *SB*, xvi, pp. 423–424.

[21] Words in brackets follow Huang's commentary.

[22] This refers to the five Confucian virtues: *jen*, *i*, *li*, *hsin*, and *chih*.

1.4 (2a3) If these two principles (law and education) are not abused, the five virtues (kindheartedness, righteousness, decorum, faithfulness, and wisdom) not abandoned, and the six tempers (likes and dislikes, cheerfulness and anger, joy and sorrow) not perverted, then the three primal spheres (*san-ts'ai*: Heaven, Earth, and Man) will be in good order, the five human activities will attain perfection,[23] the hundred officers will be in their proper places, and all meritorous [goals] of good government will be achieved.[24]

Heaven lays down the Way, the emperor sets the example, the ministers render their assistance, and the people serve as the foundation [upon which good government is built].[25]

1.5 (2b1) The government of the ancient Sage-kings consisted of: first, following the Way of Heaven; second, rectifying the self; third, employing the virtuous; fourth, taking care of the people; fifth, enlightening the institutions; and, sixth, establishing the great rule. One must accord with Heaven earnestly, rectify the self constantly, employ the virtuous permanently, take care of the people diligently, enlighten the institutions according to the canons,[26] and establish the great rule firmly. This is called the essence of government.

1.6 (2b4) The art of attaining good government consists of first, eliminating the four evils; and second, exalting the five programs of government.

1.7 (2b5) The first evil is falsehood; the second, selfishness; the third, unruliness; and the fourth, extravagance. Falsehood confounds the social conventions, selfishness corrupts the law,

---

[23] *Wu-shih*, the five human activities: demeanor, speech, seeing, hearing, and thinking; Legge, III, pp. 324–327.

[24] Legge, III, pp. 22–23.

[25] An additional couplet, 制度以綱之, 事業以紀之, "This is patterned according to regulations, and commemorated by deeds," is cited in *CSCY* 46:1a.

[26] *Tien* means precedents recorded in the historical documents.

THE SHEN-CHIEN

unruliness violates the established regulations, and extravagance destroys the proper order. If these are not eliminated, the programs of government cannot be implemented. If the social conventions are confounded, the Way will be eroded and even Heaven and Earth will not be able to preserve their natures. If the law is corrupted, the dynasty will collapse and even the emperor will not be able to preserve his rule. If the established regulations are violated, the rites will perish and even the Sage will not be able to maintain his Way. If the proper order is destroyed, desires will become reckless and [all the products of] the universe[27] will not be able to satiate their demands. These are called the four evils.

1.8 (3a8) To promote agriculture and sericulture so as to nourish life; to examine carefully one's likes and dislikes so as to rectify the social conventions; to expound the Confucian teachings[28] so as to manifest their influence; to be militarily prepared so as to maintain an august authority; and to intelligently dispense rewards and punishments in the administration of law—these are called the five programs of government.

1.9 (3b1) When people are not afraid of death, one cannot intimidate them with punishment;[29] when people find no pleasure in living, one cannot exhort them to be good. In such a case, even though we may appoint [a man as virtuous as] Hsieh to set forth the Five Lessons,[30] and [a man as meritorious

---

[27] *Ssu-piao*, the four extremities of the universe; see Legge, III, p. 15.

[28] *Wen-chiao* (*wen*, civilized, cultured; *chiao*, education, teaching), in contrast to *wu-pei* (military preparedness) in the next line, means "liberal, civil" education in general. The ancient schools of Confucianism, however, emphasized the notion of "education that transforms men" (*chiao-hua*). As expounded in 1.10 below, by *wen-chiao* and *chiao-hua*, Hsün Yüeh meant the orthodox Confucian teachings of moral conduct and the observance of rites (*li-chiao*). See also *SC* 3, note 12 on *wen*.

[29] From *Lao-tzu*, 76, tr. by Arthur Waley, *The Way and Its Power* (London, 1949), p. 234.

[30] From "The Canon of Shun," Legge, III, p. 44: "The five ranks under which human society may be ordered are: parent and child, sovereign and

as] Kao-yao to be the Minister of Crime,[31] governmental programs cannot be carried out. Therefore, the ruler must first increase the wealth of his people in order to settle their minds. The emperor must personally perform [the ritual of] ploughing the royal field; the empress must perform [the ritual of] raising silkworms.[32] There must be no idle wanderers in the empire and no wasted land in the countryside. Wealth must not be wasted in vain pursuits.[33] Corvée labor must not be arbitrarily exacted. All this facilitates sound, productive activities of the people. This is called nourishing life.

1.10 (4a5) [The ability of] a superior man to move [the heart of] Heaven and Earth, respond to the Divine Intelligence, rectify the myriad things, and achieve a kingly administration stems from his genuine sincerity. Therefore, a ruler examines carefully the regulations and follows the Way in settling his likes and dislikes. He must judge good and evil on [the basis of] merits accomplished and crimes committed; he must assess praise and blame according to a standard test;[34] he must check word against deed and carefully compare name with reality.[35] Deceit or falsehood[36] should never be allowed to dissipate the people's mind. As a result all deeds will be examined, all things

---

subject, husband and wife, brothers, and friends. *Wu-chiao* are the Five Lessons [setting forth] the duties relating to them." Hsieh, 禼 or 契, was the forefather of the Shang people and the legendary *Ssu-t'u* (Prime Minister or Minister of Instruction) at King Shun's court.

[31] From "The Canon of Shun," *ibid.* Kao-yao, 咎繇 or 皋陶, served as *Shih* (Minister of Crime) at Shun's court.

[32] The rituals were instituted by Emperor Wen of Han (179–157 B.C.) in 178 B.C. and 167 B.C. respectively. See *HS* 4:9a, 13a; H. H. Dubs, I, pp. 242, 254. The practices had their origin in the teachings of the Agriculturalist School, which had been severely criticized by Mencius; Legge, II, pp. 246–257.

[33] 虛用, "used in vain," reads 賈用, "used in trading," in the quotation of *SC* in *HHC* 29:15a. Commercial activities at that time were considered non-productive. My translation reconciles both versions.

[34] Legge, I, p. 301.

[35] *Ibid.*, pp. 263–264, 300–301.

[36] 詐偽 reads 詐偽淫巧 (deceitful, false, lewd, tricky) in *HHC* 29:15b and *CSCY* 46:2a.

will be successful, all goodness will be made known, and all evil will be openly denounced. In social conventions, there will be nothing licentious or strange; among the people, there will be no lewd habits.

When the people of the hundred surnames observe that benefit or injury lies within oneself, they will hold their minds in solemn respectfulness and regulate their conduct conscientiously. Inwardly there will be no excesses or bewilderment; outwardly they will have no improper ambitions, lest these be known. Those who commit offences will have no chance to escape; those who are innocent will have no worry in their minds.[37] Illicit requests and improper visits should not be accepted; bribery should not be practiced. In this way, the people's minds will be set at peace. This is called rectifying the social conventions.

1.11 (4b7) A superior man can be moved by [noble] feeling; a petty man can be controlled by punishment. Honor and humiliation constitute the essence of reward and punishment. The teaching of rites and [the ideas of] honor and humiliation are used with respect to a superior man in order to cultivate his [noble] feelings; the fetter and handcuff together with the whip and the stick are used upon a petty man to administer punishment. A superior man will not incur humiliation upon himself, let alone statutory punishment; a petty man is not afraid of punishment, let alone humiliation. But for an ordinary man, both statutory punishment and the teaching of rites must be used.[38] In the absence of proper teaching, an ordinary man may be pushed down into the ranks of petty men; when proper teaching has become prevalent, he may be led on to the Way of a superior man. This is called manifesting the influence [of Confucian education].

[37] The *SPTK* edition reads: 去徼倖, 無罪過, 不憂懼, "There will be no opportunist, no crime, no fear." This appears to be somewhat out of line with the preceding sentence. My translation follows the quotation in *CSCY* 46:2b.

[38] *Chung-jen*, a man who stands between the two extremes, i.e., between the superior man and the petty man. For the three personality types, see *SC* 3.7, 5.15.

1.12 (5a7) The feelings of a petty man are such that when you treat him easily, he becomes haughty; being haughty, he becomes licentious; being licentious, he becomes restless; being restless, he becomes spiteful; being spiteful, he becomes rebellious. In times of danger he becomes subversive; and in times of peace he gives rein to his desires.[39] If there is no august authority, nothing can restrain him. Therefore, a ruler must remain militarily prepared to guard against unexpected troubles and suppress riotous outlaws. In times of peace [a program of military training] may be entrusted to the civil authority,[40] and in times of crisis [this can provide] the fighting units for military campaigns. This is called maintaining an august authority.[41]

1.13 (5b7) Reward and punishment are the pivots of government. Make rewards clear and punishments certain. Promises must be made with care, and orders must be issued with discretion. Rewards must be dispensed to encourage the good, punishments to correct the bad. A ruler must not reward indiscriminately—it is not that money is too precious [to be used for such a purpose], but that once indiscriminate rewards are made, good behavior will not be encouraged. He must not punish inappropriately—it is not only that he is sympathetic with men [suffering such punishments], but that once inappropriate punishments are made, evil-doers will not be corrected.

Not to encourage good behavior is to stop people from doing good; not to correct evil-doing is to allow people to do evil. If a ruler is capable of not stopping his people from doing good and of not allowing his people to do evil, the state will be in good order.[42] This is called the administration of law.

[39] Cf. quotation of *SC* in *CSCY* 46:3a.
[40] The policy was attributed to Kuan Chung (d. 645 B.C.), who served Duke Huan of Ch'i. *Kuo-yü* (*SPTK* ed.) 6:6–7a; *Kuan-tzu* (*SPTK* ed.) 8:7b–8.
[41] For Hsün Yüeh's discussion of this institution, see *SC* 2.6.
[42] Translation follows the quotation of *SC* in *HHS* 62:13b, ending with the clause 則國法立矣. The *SC* quotation in the commentary of the *Ti-fan* (Kuang-ya shu-chü ed.) 3:8a, has three additional lines: 刑不濫而威立矣；賞不僭

THE SHEN-CHIEN

1.14 (6b5) Once the four evils are eliminated and the five programs of government are implemented, the ruler should follow this path sincerely and adhere to it without wavering. He is simple but not negligent, he is lax but does not err, and he rules by "non-action" so that his subjects can govern themselves and think for themselves.[43] Then the state will be in good order and the people will be transformed without much coercion.[44] [If] the ruler merely wears his robes, folds his hands, bows, and shows respect [to the virtuous], all within the four seas will be pacified.[45] This is called the art of attaining good government.

1.15 (6b2) There are six principles [that a ruler must observe] in order to establish the pattern of the Way:[46] first, the golden mean, *chung*; second, harmony, *ho*; third, justice, *cheng*; fourth, public-spiritedness, *kung*; fifth, sincerity, *ch'eng*; and sixth, comprehensiveness, *t'ung*. From the Way of Heaven comes the golden mean; from the Way of Earth comes harmony; from the virtue of humanity comes justice; from the myriad activities comes public-spiritedness; from the exemplar of oneself comes sincerity; from the changing lots[47] comes comprehensiveness. These are called the substance of the Way.

而化行矣；既不僭不濫，則為惡者知所懼而為善者知所觀矣. Since it is not clear whether these are quotations from the *SC* or words added by the commentator of the *Ti-fan*, I have refrained from incorporating them into the text.

[43] Cf. *CSCY* 46:3b. From *Lao-tzu*, 48; see A. Waley, *The Way and Its Power*, p. 201.

[44] Translation follows the version in *HHS* 62, which has the additional line, "不嚴而化."

[45] See Legge, III, p. 316, alluding to the legendary Sage-kings Yao and Shun.

[46] *Tao-ching*, warp of the Way, as compared with *tao-shih*, contents or fruits of the Way.

[47] *Pien-shu*, changing number (lot, condition); cf. *I-ching* 7:13b. According to ancient Chinese cosmological-numerological thinking, the cosmos consists of a binary *yin-yang* structure of even and odd numbers, represented by the broken (– –) and unbroken (—) lines of the symbolic trigrams and hexagrams in the *Book of Changes*. It was believed that an understanding of these symbols and the numerological principles behind them could lead to comprehension of both the eternal essence and the changing conditions of nature and the cosmos.

1.16 (6b6) There are ten difficulties that a ruler must consider in employing the virtuous and talented: first, not recognizing them; second, not elevating them; third, not trusting them; fourth, not [continuing to trust them] to the end; fifth, denying great virtue because of a minor offense; sixth, disregarding great merit because of a minor mistake; seventh, disregarding outstanding excellence because of minor fault; eighth, injuring the loyal and upright because of evil slander; ninth, betraying correct principles because of incorrect theories; and tenth, deserting the virtuous and talented because of jealous stricture. These are called the ten difficulties. If they are not overcome, the virtuous ministers cannot be employed. When a ruler cannot employ the virtuous ministers, the state will not be his.[48]

1.17 (7a5) A ruler should examine the nine conditions of a state in order to determine the principles of his government. The first condition is order; the second, decline; the third, weakness; the fourth, perversion; the fifth, disorder; the sixth, ruin; the seventh, rebellion; the eighth, danger; and the ninth, extinction.

1.18 (7a7) The state is in an orderly condition when the ruler and the ministers behave cordially and decorously toward each other, when the hundred officials are in harmony without conforming blindly, yielding and not contentious, hardworking and uncomplaining, and each of them routinely discharges his own duties in office.

1.19 (7a9) The state is in a condition of decline when the rites and conventions are not uniform, the government offices and duties are taken lightly, the petty officials indulge in jealous slander, and the commoners discuss matters irresponsibly.[49]

1.20 (7b2) The state is in a weakened condition when the

---

[48] Collation from *CSCY* 46:3b–4a.
[49] Cf. Legge, I, p. 310.

THE SHEN-CHIEN

ruler is submissive,⁵⁰ the ministers are remiss, the literati are complacent, and the masses are without direction.

1.21 (7b4) The state is in a perverted condition when the ruler and minister contend with each other for prestige,⁵¹ the imperial court seeks improper favors, the literati compete for social glamor, and the commoners strive for profit.

1.22 (7b5) The state is in a disorderly condition when the ruler has many desires and those below him raise many issues, and when the laws are not fixed and administration is handled by divergent authorities.

1.23 (7b6) The state is in a condition of ruin when extravagance is mistaken for generosity, eccentric nonconformity for loftiness, and lack of principle for versatility, or when the observance of rites is considered burdensome and the obedience of laws is considered too restrictive.

1.24 (7b8) The state is in a condition of rebellion when acrimony is mistaken for efficiency and profit-seeking for public-spiritedness, and when those who oppress their subordinates are considered capable and those who [blindly] comply with their superiors are considered loyal.

1.25 (8a1) The state is in a dangerous condition when the ruler and his subjects become estranged, the inner and outer factions⁵² deceive each other, the petty officials compete for illicit favors, and the high-ranking ministers struggle for power.

1.26 (8a3) The state approaches extinction when the ruler

---

⁵⁰ *Jang* 讓 (yielding) reads *ch'ien* 謙 (humble) in *CSCY* 46:4a. Another reading, *hsien* 嫌 (suspicious), was recorded in Huang's commentary.

⁵¹ *Ming* 名 (name) reads *ming* 明 (intellect, clear comprehension) in *CSCY* 46:4b, and *meng* 盟 (alliance, covenant) in *Hsiao-Hsün-tzu* 4a.

⁵² Probably referring to the ruling family and the governmental establishment, or the palace and the court.

THE SHEN-CHIEN

does not consult others and his subjects do not remonstrate, and when the ruler heeds the words of [his favorite] women, and illicit personal rule prevails. Therefore a ruler must examine the condition of his state.

1.27 (8a6) Serious consideration must be given to the use of statutory punishment on the masses so as to manifest human feelings.[53] The greatest virtue of Heaven and Earth is [the giving and sustaining of] life.[54] The ultimate end of the myriad things is death.[55] The dead cannot come to life again, and the one who suffers punishment by mutilation can never recover [from such injuries].[56] Therefore, when the ancient Sage-king inflicted a punishment he would have the various officials discuss the charges, set up court hearings at different [administrative] levels to reach a verdict, carefully consider the human feelings [involved] to moderate the sentence accordingly, and then have the punishment announced in court and carried out in the marketplace.[57] And still he would show sympathy for the criminal and condole [with his family]. On the day of the

---

[53] The sentence alludes to the famous Chinese legal term *ch'ing-hsin* 情訊, which appears in the chapter on judicial officers in *Chou-li* 35:2a, 3a. It means that in a legal investigation careful consideration should be given not only to external evidence but also to the inner motives of the offender, so that the whole picture may be correctly apprehended. The statement may also refer to Confucius' ideal of legal prosecution, regarding which he was recorded as saying that in order for litigations to cease, those who are not honest (or who do not stand on the foundation of truth, 無情, *wu-ch'ing*) must find it morally impossible to testify. See Legge, I, p. 364.

[54] *I-ching* 8:3b, Legge, *SB*, XVI, p. 381. This sentence is frequently quoted by Chinese officials in their discussions of the legal system.

[55] Alluding to the first of the six misfortunes mentioned in "the Great Plan"; Legge, IV, p. 343.

[56] For a brief discussion of the abolition of punishment by mutilation under the reign of Emperor Wen of Han, see H. H. Dubs, I, p. 255. See also A. F. P. Hulsewé, *Remnants of Han Law*, I (Leiden, 1955), pp. 124–128.

[57] A very meticulous judicial procedure is described in the two chapters on offices of judicature in *Chou-li* 35:2a–23b; 36:1a–3a, 11b–13b. Although the authenticity of this book is now generally suspect, Hsün Yüeh tended to accept its descriptions of the exemplary institutions of the Chou as true.

execution music would not be played. How much attention must be given to such matters! Oh, how careful one must be [in the infliction] of statutory punishment!

1.28 (9a5) Consideration must be given to the five occasions for pardon so as to put the people's minds at ease.[58] [Leniency is possible] first, after consideration of the original motive [of the accused]; second, because of the past virtues [of the offender]; third, as a means of encouraging the meritorious; fourth, in order to exalt moral education;[59] and fifth, as a measure of expediency.[60] The pardons ordered by the former kings had to be made on the basis of one of these conditions. Otherwise punishment should be imposed without pardon.

1.29 (9a9) The son of Heaven divides his day into four periods: in the morning, he attends to affairs of state; in the daytime, he makes inquiries and consultations; in the evening, he reviews the ordinances; at night, he gives rest to his body.[61] In the upper ranks there are royal tutors; in the lower ranks, there are attending officials.[62] Major problems require study and minor ones consultation. A ruler should not reject straightforward and honest counsel; nor be ashamed to ask and learn from his inferiors;[63] nor confuse his official and his private

---

[58] For a detailed discussion of amnesty in Han times, see Hulsewé, I, pp. 251–279; also T'ung-tzu Ch'ü, *Law and Society in Traditional China* (Paris, 1961), pp. 41–87.

[59] Compare this with the considerations for amnesty mentioned in *Chou-li* 35:4. See also Hulsewé, I, pp. 214–224, 285–298.

[60] Note the compromise stand taken by Hsün Yüeh on a number of critical issues under the pretext of "expedience."

[61] This is a quotation from the advice given by Kung-sun Ch'iao to the head of the state of Chin upon the illness of the latter in 540 B.C.; see Legge, V, p. 573–580.

[62] According to *Shu-ching*, the Chou king had three senior and three junior advisors to serve him personally, i.e. the Grand Tutor, the Grand Assistant, and the Grand Guardian, and the Junior Tutor, the Junior Assistant, and the Junior Guardian. See Legge, III, pp. 426–528.

[63] From *Confucian Analects*, Legge, I, p. 178.

THE SHEN-CHIEN

[business].[64] Also, he should not act privately and publicly according to different standards. This is called communication [between the high and the low].

1.30 (9b8) [Answer to] a question:[65]
One who enlightens his rule begins with that which is close at hand. The foundation of the myriad things lies in oneself; the foundation of the empire lies in the family; the foundation of good or bad government lies in the left and the right [hands of the ruler]. If [the ruler] stands upright at the center, the country in the four directions will be in good order.

1.31 (10a2) [Answer to] a question:
He who comprehends the Way holds to a simple principle.[66] There is one word which may be constantly observed—"sympathy" (*shu* or reciprocity).[67] There is one path which one may constantly walk—"justice" (*cheng* or uprightness). Sympathy is the art of kindheartedness; justice is the essence of righteousness. Oh, how far-reaching! This is called the root of the

---

[64] Under the Han dynasty, a tenuous distinction was made between the affairs of state and the affairs of the ruling house, both of which were headed by the emperor. For instance, the court assembly (known as the "outer court" under the Former Han) was differentiated from the palace assembly (known as the "inner court"); and there was a *Ta-ssu-nung* (Grand Minister of Agriculture) in charge of the financial affairs of the state and a *Shao-fu* (Small Treasurer) in charge of the financial affairs of the ruling house. See Yu-ch'üan Wang, "An Outline of the Central Government of the Former Han Dynasty," *HJAS* 12 (1949) 155, 161–178.

[65] The seven characters following the word *wen* may constitute the concluding part of the question or the beginning part of the answer. In either case, the sentence appears to be incomplete. This may be due to a corruption of the text or to Hsün Yüeh's peculiar style of introducing a dialogue with the isolated character *wen* while omitting the content of that question. Since the same expression appears six times in *SC* 1 (1.30, 31, 33, 34, 35, 36), the second explanation seems more likely.

[66] Legge, II, pp. 494–495.

[67] Legge, I, p. 301, rendered the word *shu* as "reciprocity" in accordance with Confucius' own remark: "What you do not want done to yourself, do not do to others."

Way. All the myriad changes lie therein. This is called attainment without special deliberation and accomplishment without much ado. Simply keep this [principle] in mind and your merit will be as great as the universe.

1.32 (10b1) From the son of Heaven to the commoners, the range of likes and dislikes, of sadness and joy, is the same. But [the gradations between] sumptuous and frugal, laborious and leisurely [life-styles] must be regulated. The upper classes should have sufficient means to perform the rites; the lower classes should have sufficient means to enjoy [life].[68] This is called the Great Way.

The state and all under Heaven are a whole body. The sovereign is the head; the ministers are the limbs;[69] the people are the hands and feet. When below there are people grieving, the one on high should not have full-scale musical entertainment;[70] when below there are people starving, the one on high should not order a full meal; when below there are people who are cold, the one on high should not put on full dress. To wear a crown with hanging fringes of pearls while [standing] in bare feet is not proper. Cold feet hurt the heart; people who are cold hurt the state.

1.33 (10b9) [Answer to] a question:

The sovereign leads his people to the best way; the people nourish their sovereign with the best things. The sovereign bestows his favour; the people present their achievements. There is no going without a return—this is the principle of reciprocity.[71] Although means [of entertainment] are employed full-scale during times of great peace, this is not done in order to satiate one's desires; and although the performance of rites

---

[68] *Yüeh* (*lo*) means either music or joy or a combination of these two, "entertainment." Here the rendering follows the context of *SC* 1.9.

[69] Legge, III, pp. 79, 89, 90.

[70] Here again the meaning of *yüeh* (*lo*) is not clear; see note 68.

[71] Cf. note 81. For further discussion of the concept of *pao* (reciprocity), see *Hsün Yüeh*, pp. 54–56.

is reduced when means are insufficient, this is not done merely to be modest and frugal. For these things are prescribed by the principles of numerology.[72]

1.34 (11a4) [Answer to] a question:
A ruler may impose official taxes but not private levies; he may make public expenditures but not have private expenses; he may exact official corvée but not private services; he may bestow official rewards but not personal favors; he may display public indignation but not private grievances.

If private levies are imposed, those below will be troubled improperly; this is called doing damage to the pure. If private expenses are provided, the official expenditures will be boundless; this is called doing damage to the proper institutions. If private services are exacted, the people will be disturbed and upset without limit; this is called doing damage to righteousness. If personal favors are bestowed, those below will harbor vain expectations without proper guidelines; this is called doing damage to what is just. If personal grievance is displayed, those below will become suspicious, fearful, and unsettled; this is called doing damage to virtue.

1.35 (11b2) [Answer to] a question: one who governs his people well, governs their nature.

Someone said: "Melt metal and it liquifies, but remove the fire and it solidifies [again]; pump water and it rises, but leave it alone and it comes down [again]. How can you regulate these things?"

I said: "Never remove the fire and [the metal] will remain liquified; never stop pumping and [the water] will keep rising. The great metallurgical brazier can prevent [the metal] from

---

[72] *Shu* 數 (number, lot, fate) means the "mechanism," interpreted by some scholars of the *I-ching* as the principle of synchronicity which regulates the correspondence between conditions in the cosmos, in nature, and among men. See note 47 above. For further discussion of Hsün Yüeh's ideas of number-mysticism, cosmic cycles, reciprocity, and fate, see *Hsün Yüeh*, pp. 106–107, 138–141.

THE SHEN-CHIEN

solidifying; the water pump can prevent [the water] from going down. If a good educator acts in this way, throughout his life [his people's natures] will be well-governed. Thus an ordinary man may be made to follow the path of Yen [Hui] and Jan [Po-niu].[73]

"Throw one hundred [pieces of] gold in front [of a man] and at the same time press the naked blade of a sword against his body; then even [a man as bold as] the giant Chih[74] will not dare to pick them up. If a good legislator acts in such a way throughout his life, nobody will dare to pick up the gold. Therefore, even [a man as bad as] Chih may be made as virtuous as Po-i."[75]

1.36 (12a6) [Answer to] a question:

People are like water. To cross a great river, the best way is [to go] by boat and the next best is to swim. Swimming is toilsome as well as dangerous, whereas riding in a boat is comfortable and safe. If one goes vainly into the water, [unable to swim and without a boat], he is certain to drown. Governing the people with wisdom and skill is like swimming; governing the people with grace and virtue is like riding in a boat. Giving rein to the people's passions is called [governing in an] unruly manner; [trying to] exterminate the people's passions is called ruining [the state].

Someone asked: "Then what should we do?"

I said: "Set a limit and do not let them transgress [it]. Also, make room for them so that they will not transgress. You can stop water from overflowing, but you cannot stop it from flowing."

---

[73] Both were exemplary disciples of Confucius; see Legge, I, p. 237.

[74] Chih, a notorious brigand in Confucius' time; *Shih-chi* 61:8b, tr. by B. Watson in *Ssu-ma Ch'ien, Grand Historian of China* (Columbia University Press, 1958), p. 189.

[75] Po-i, an exemplary recluse in ancient China, who would not eat the food grown by the rebelling Chou people and thus starved himself to death. See *Shih-chi* 61:7–8a; B. Watson, *Ssu-ma Ch'ien*, pp. 187–188.

THE SHEN-CHIEN

One who knows well how to impose restriction restricts himself first, and then others. One who knows not restricts others first, and then himself. One who knows exceedingly well how to impose restriction may govern virtually without restriction. The administrative ordinances should also be like this. If one allows his own passions to be free but restrains the desires of the masses, acts deceitfully in office but exhorts his people to be honest, seeks what he already possesses in excess and deprives those below of what they still need, and leaves what is easy for himself but presses hard for the difficult [tasks] to be done by others, then grievances will arise. This is to cut off the source of what is called reasonable administration.[76]

1.37 (13a1) For those in high positions to control those below, it is like [practicing] the art of the fisherman: he hides [the pole] in his hands, makes the hook respond [from a distance], and thereby catches the fish. For those nearby to control those at a distance, it is like driving a horse: [he takes the reins] lightly in hand and controls the bit [from a distance], thus guiding the horse. Therefore, if a Way does not reach oneself it is not a [true] Way.[77]

Look at a small boy herding chickens and you will learn a lessons about governing the people. When the boy presses the chickens too hard, the chickens become alarmed; when he herds them too gently, the chickens become unresponsive. If the chickens are going north and are stopped suddenly, they abruptly turn south; if they are heading south and are stopped, then they abruptly turn north. When the boy comes too close, they fly up; when he lets them loose, they are remiss. When their minds are at ease, you can approach them; when they stagger and become unsettled, you should feed them. The best way of herding is by not herding. When their minds are settled, they will follow the way and go indoors.

---

[76] *Li* 理, reason of state, or to administer.
[77] Cf. Legge, I, pp. 393–394, 396.

THE SHEN-CHIEN

1.38 (13b4) The best way is not to "trade in vain" (or to make false promises);[78] next, not to commit theft; next, not to commit robbery. A person in a high position pacifies those below by deeds and favors; those below support a person in a high position with money and labor. In this way, the high and the low stand by each other.[79] If he "trades in vain," the people will not stand by [him]; if the people do not stand by [him], the ruler will get it through deceitful means—this is called theft. If he commits theft, the people will prepare for this; when the people prepare for this and he cannot get it [through deceitful means], he will press the people hard and grab it; this is called robbery. In such a case, the people will fight back and there will be disasters and riots.

1.39 (14a1) Someone asked: "Does the Sage-king take the empire of all under Heaven as a [thing of] pleasure?"

I said: "No, he takes it as a concern. [In such a case, the people of] all under Heaven regard the Sage-king [as the cause of] their happiness. A mediocre ruler takes the empire of all

---

[78] The meaning of *k'ung-shih* 空市 is not clear. It may mean "vain trade," or "exhaust the market." The character 空 also may be a corruption of 罔 *wang*. In Mencius, Legge, II, pp. 227–228, a passage reads: "In times of old, the market-dealers traded goods... and there were simply certain officers to keep order among them. It happened that there was a mean fellow, ... who caught in his net all of the market's profits 罔市利. The people therefore proceeded to place a tax upon his wares. The taxing of traders arose from [the bad conduct of] this mean fellow." The precept of not taxing trade was stated earlier in *ibid.*, p. 162, and reiterated on pp. 199–200. See also *Han-shih wai-chuan*, tr. by J. R. Hightower (Harvard University Press, 1952), p. 116. The ideal of not taxing trade is in basic accord with Hsün Yüeh's liberal policy towards commerce. However, in *Kuan-tzu* 12:9a there is an interesting discussion of the relation (mutual aid, 與 *yü*) between the court and the peasants reading: "to *shih* 市 (trade) is to *ch'uan* (exhort)." The annotation states: "The metropolitan people (*shih-jen*) do not vainly take 虛取 (*hsü-ch'ü*); therefore the country people do not vainly give (*hsü-yü*)." Hence *K'ung-shih*, to vainly trade or to vainly exhort (or make vain promises).

[79] *Shang-hsia hsiang-yü* means that the high and the low give and take from each other, or support each other. See note 78.

under Heaven as a [thing of] pleasure. [In such a case, the people of] all under Heaven regard him with concern. The Sage-king constrains himself so as to expand the happiness of his people; the mediocre ruler expands himself and constrains his people to the point of grievance. Expanding the happiness of his people, happiness will also be his reward; constraining his people to the point of grievance, grievances will also afflict him.[80] This is the Way of all under Heaven."

1.40 (14b1) In times of good order, a ruling post is highly regarded for three [reasons]: first, [as a means of] extending the Way to all under Heaven; second, [as a means of] extending favors to the people; and third, [as a means of] enhancing one's own virtue. In times of decadence, a ruling post is prized for three [reasons]: first, as the high position from which to impose on others; second, [as a means of] enjoying oneself, using wealth [stolen from the state]; third, [as a means of] satiating one's desire for revenge.[81]

The ruling post in times of good order is a true ruling post, whereas in times of decadence it will only beget trouble. If one improperly imposes on others from his high position, others will injure him; this will cause trouble. If one improperly enjoys himself, he will be deprived of his enjoyment; this will cause trouble. If one improperly satiates his desires, the reverse will come of it; this will cause trouble.

1.41 (14b8) In times of good order, a minister is highly regarded for his submission to three things: first, to the mind (intellect, conscience, etc.); second, to his duty; and third, to

---

[80] Cf. Legge, II, pp. 157–158.

[81] *Pao* may mean to enjoy, to reward, or to avenge. Translation follows Huang's commentary. See L. S. Yang's article, "The Concept of *Pao* as a Basis for Social Relations in China," in John K. Fairbank, ed., *Chinese Thought and Institutions* (University of Chicago Press, 1957), pp. 291–309; also Yang's review of A. Wright, *Buddhism in Chinese History*, in *Harvard Journal of Asiatic Studies* 23 (1961), 216.

the Way. [In times of decadence, a minister is highly regarded for his submission to three things: first, to his body (desires, lusts, etc.); second, to words (vain talk, pretense, etc.); and third, to events (chance, outside pressures, etc.).][82]

Submission in times of good order is true submission, whereas in times of decadence it only begets perversion. Improper submission to the body perverts one's temper; improper submission to words[83] perverts one's sense of loyalty; improper submission to events perverts the Way.

1.42 (15a3) When the high and the low are not in proper order, government office is slighted; when official ranks and grades are not firmly established, government office is slighted; when emoluments are set too low and those in inferior positions receive undue favors, government office is slighted; when [the structure of the] administrative bureaus and their duties are repeatedly changed, government office is slighted; when the transfer of officials becomes confusingly and annoyingly frequent, government office is slighted; when promotions and demotions are not intelligently made, government office is slighted; and when [the ruler] does not treat his ministers according to the [proper] rites, government office is slighted. There has never been a case in which an office that is slighted can do anything weighty. The most precious possession of a Sage is his office.[84] When his office is slighted, he loses what is most precious to him.

1.43 (15b1) Oh, the likes and dislikes [of the ruler] have for a long time failed to exert an influence upon the conventions of the vulgar.[85] [One may] exalt moral integrity, [but] slip

---

[82] Collation by Huang Hsing-tseng; cf. also *CSCY* 46:5b.
[83] Translation follows the quotation of *SC* in *CSCY*.
[84] From *I-ching* 8:4a; see Legge, *SB*, XVI, p. 381.
[85] The passage may also read: "When the likes and dislikes (of the Sage) are not manifested, vulgar customs will rise to the fore."

easily into narrow-minded and unrefined stubbornness.⁸⁶ [One may] hate arrogant authority [but] admire the powerful. [One may] despise covetous desires [but] exalt efficiency in [achieving profitable] gains. [One may] honor self-improvement [but] prize praise from others. The myriad things are all like this.

Mind and words, words and deeds—these three are correlated. Likes and dislikes, praise and blame, reward and punishment—these three also act reciprocally.⁸⁷ If these six malfunction, reality becomes confused. Those who adhere to reality will become more prosperous; those who seek self-improvement will become more successful in the search; those who live in darkness will shine more brightly. In this way, the people will come to understand what is essential.

---

⁸⁶This sentence and the three that follow appear to be ambivalent and illogical. For instance, the second sentence, lit. "to dislike the powerful authority and to honor the powerful authority," is absurd. It might mean either that the Sage dislikes the one but the vulgar honors the other, or that, when likes and dislikes are confused, the vulgar will dislike the one but at the same time honor the other. See Hsün Yüeh's discussion of likes and dislikes in *SC* 1.10.

⁸⁷福 *fu* should read 福 or 副; Lu, *Corrigenda*, 1a.

Shen-chien 2
Current Affairs
(Shih-shih)

2.1 (1a4) The most important [affairs of the day can be divided into] twenty-one categories. The first two categories are exalting knowledge and prizing honesty. The remainder,[1] altogether nineteen items, are also crucial and ought to be dealt with [by the sovereign]. Of these, the first concerns [the establishment of] enlightened merit examinations [for officials];[2] the second, that the senior and junior ministers (*kung* and *ch'ing*) should not feel reluctant[3] to receive appointments in the commandery offices, and officials of the rank of two thousand piculs[4] should not feel reluctant to accept appointments in the district governments; the third, establishment of an office to exalt the military [arts]; the fourth, discussion about the Shepherd of the Province (*Chou-mu*); the fifth, [saving] the life [of those who would otherwise] die [as a result of receiving the death penalty] by substituting punishment by mutilation; the sixth, use of both moral influence and statutory punishment; the seventh, establishment of a rule for avoiding vengeance; the eighth, discussion about

---

[1] Lit. "the second ones."

[2] By *K'ao-shih* (examinations) Hsün Yüeh did not mean the full-fledged civil service examination system, which was developed from the seventh century onward and became firmly established only after the tenth century. What he was referring to here was the system of merit-promotion for officials, as described in *SC* 2.4.

[3] *Chü*, restrictive. As elaborated in *SC* 2.5, however, the issue was not that the court or high official making such appointments was restricted by any regulations or traditions, but rather that for political or personal reasons most prominent scholar-officials refused to accept the appointments.

[4] Mostly governors of the commanderies.

official emoluments; the ninth, discussion about absolute [ownership of] land; the tenth, discussion about the monetary system; the eleventh, simplification of sacrifices and maintenance of the important ones; the twelfth, correspondence between Heaven and Man; the thirteenth, court audience on the first day of the month; the fourteenth, exalting [orthodox] teachings inside [the palace]; the fifteenth, appointment of Official Erudites; the sixteenth, the important way of cultivating the highest standards of morality; the seventeenth, prohibition of the frequent [granting of] amnesty; the eighteenth, rectification of the practice of *shang-chu*;[5] the nineteenth, reinstitution of the Outer and Inner Record-keepers.[6]

2.2 (1b5) [When] P'an-keng moved [his capital] to Yin, he put a stop to extravagance and practiced frugality.[7] He transformed and refashioned the customs, making adjustments in accordance with the times. Now plentiful, now scarce, now full, now empty—[as conditions change], one cannot always follow a single way.[8] Exalting knowledge and prizing honesty are the methods [of coping with change appropriate] for both the past and the present.

When the population is sparse, supplies will be easy to obtain; when land is spacious, things will be easy to produce;

---

[5] *Shang-chu*, lit., serving the princess; i.e., the husband of the princess is treated as the inferior partner in the matrimonial relationship.

[6] The archivists of the inner court (the emperor's household chambers) and the outer court (formal court bureaucracy), as elaborated in *SC* 2.22.

[7] P'an-keng, the nineteenth ruler of the Shang dynasty, reportedly made several important pronouncements concerning the moving of the Shang capital to Yin ca. 1401 B.C. See Legge, II, pp. 220–247.

[8] These two sentences sum up the important teachings of the *I-ching* and *Lao-tzu*. For the term *ying-hsü* (fullness-emptiness), see A. Waley, *The Way*, pp. 146–152. The sentence may also refer to P'an-keng, since in one of his purported statements he proclaimed that "the former kings reverently obeyed the Mandate of Heaven, but still they did not enjoy constant repose and did not perpetually abide in just one city" (Legge, III, pp. 222–223). However, it appears more likely that the subject of the sentence in the *SC* quotation is not P'an-keng himself but "the former kings" mentioned in his statement.

when work is simple, businesses will be easy to establish. Tired of disorder, one will think about order; having experienced the difficulty [of action],[9] one will think of quietude.

2.3 (2a2) Someone said: "[During the period of] the Three Kings,[10] the people were most honest and the government was very pure. Was this because of [their] Heaven-granted natures?"

I said: "The people under the Three Kings were honest, whereas the people under the Ch'in were corrupt. [This is due to the different conditions of] the times. Mountain-dwellers are simple, whereas city-dwellers are sophisticated. [This is because of the difference between their] environments. King Chieh and King Chou did not change their people,[11] and disorder and rebellions occurred. The people under King T'ang and King Wu were no different,[12] but orderliness and peace prevailed. [This is because of the difference between the two] governments. The population under the Three Kings was sparse. And this sparseness [of population] enabled the people [to stay] honest. The policies [followed by] the Three Kings were pure, and this purity made [their administration] incorruptible. Was this only [due to] human nature?[13] When [rulers] do not seek things of no [real] worth, do not

---

[9] *Ch'uang-nan* may also mean "the difficulty of founding [a new dynasty]."

[10] Legendary rulers of remote antiquity. There are different accounts of who these three Sage-kings were: i.e., T'ien-huang, Ti-huang, and Jen-huang; Fu-hsi, Shen-nung, and Huang-ti; or Yao, Shun, and Yü.

[11] Chieh and Chou, last rulers of the Hsia and Shang dynasties. *I-min* may mean "to have a different type of people" or "to make their people change (to reform them)." My translation combines both meanings. Cf. Legge, II, p. 410.

[12] King T'ang and King Wu, acclaimed founders of the Shang and Chou dynasties respectively. See note 11.

[13] *Hsi* 奚 reads *i* 矣 in *CSCY* 46:6a, whereupon the character becomes the ending particle of the preceding sentence, while the character *hsing* (human nature) becomes the subject of the following sentence. However, what is described in the following sentence is not human nature (*hsing*) but human efforts. My translation follows the first version.

accumulate goods that are difficult to procure, eliminate glamorous and luxurious decorations, and keep off the path of profit-seeking and self-advancement, then the social customs will be purified. Cut out trivial taboos, discontinue unorthodox cults, and do away with strange superstitions; then apparitions and falsehood will cease to exist. If [the ruler] sets his mind on sincere devotion, relies on self [-improvement], and rectifies matters of great importance, then even the Divine Intelligence will respond [favorably] to him. Ban heretical teachings, get rid of improper levity in thinking, suppress the hundred schools of philosophers, and exalt the sacred orthodoxy—then the Way will be established. Put aside frivolous and flowery matters, uphold things of real merit, eliminate unproductive pursuits, and identify with the fundamental;[14] then [great] deeds will be accomplished."

2.4 (2b5) Whenever criticism or praise is offered, there should be a means of testing its validity.[15] [This] is a standard by which all things should be assessed. One who is selected for a specific [task] should be tested in that matter. One who is in charge of a specific office should be tested by the deeds he accomplishes [in that office]. If rewards and punishments are made upon real grounds, how can [evil] ever reverse them?[16] How can a man conceal [what he has done]?

There is a saying: "Even Chih the brigand[17] could not steal a foot or an inch of land." If an inch [of land] could not be stolen, how much more unlikely would it be for a foot [of land to be stolen]! If the examination of all other matters

---

[14] *T'ung* 同 reads *chou* 周 in *CSCY* 46:6a.

[15] From the *Confucian Analects*, Legge, I, p. 301.

[16] In the Ch'ung-wen shu-chü 1875 edition and the *SPPY* edition the phrase reads 賞罰失實, "If rewards and punishments are not made upon real grounds, [evil will reverse them]." However, 惡 (evil) may also read *wu* 惡 (how). The interrogative, "How can these rewards and punishments ever be controverted?" together with the next interrogative, "How can a man conceal what he has done?" have a strong, positive connotation, in contrast to the *SPPY* version.

[17] See *SC* 1.35, note 74.

were like the measuring of the paddy-fields which stretch across the open countryside, then those who practice selfishness would be few in number. If it were like a boat that can shoot the rapids but sinks on the bank of a harbor, then its credibility would be destroyed. Therefore, [state] affairs must be examined on their merits; words must be examined for their practicability; those in active life must be examined on their conduct; those in repose must be examined according to [the ideals] to which they adhere.[18]

2.5 (3a6) The senior and junior ministers[19] are not customarily appointed as administrators of the commanderies; the officials of the rank of two thousand piculs[20] are not customarily appointed as magistrates of the districts. [This custom] is not proper. The minor [officials] who have proven themselves competent in office should ultimately be promoted to high-ranking posts; then those in lower ranks will nurture a competitive spirit. The senior officials who have failed in their assigned duties should be demoted to lower posts; then those in high ranks will be more prudent. When the sacred tripod overturns, the [broken leg which was the casue of the overturning] should be liable to statutory punishment.[21] Why should [those in authority] be reluctant to demote [such officials]? If [those who have] the ability to govern [a region of] one thousand *li*[22] cannot be induced to fill [the vacant positions in] the commanderies, and [consequently] the administrative work of the district towns is neglected, how unfortunate is this! Those who are dismissed for some reason other than their own fault should not be demoted—

---

[18] The whole passage explains what Hsün Yüeh meant by "enlightened examinations" in *SC* 2.1.

[19] *Kung* (Ducal Ministers) and *ch'ing* (the nine ranking ministers).

[20] See note 4.

[21] From *I-ching* 5:22b, Legge, *SB*, xvi, p. 170; see also the translation by Richard Wilhelm, rendered into English by Cary F. Baynes (New York, 1950), p. 209. *Kung* here signifies the senior minister, not the ruler or the prince (duke) as rendered by Legge and Wilhelm.

[22] Administrators of the provinces, *chou*, or commanderies, *chün*.

this is to treat the worthy officials with respect. Those who lose their posts because of their own fault should be demoted—this is to punish their misconduct.[23]

2.6 (3b8) Emperor Hsiao-wu (r. 140–87 B.C), taking account of the fact that the barbarians of the four borders had not yet submitted themselves [to the imperial court] and that bandits and outlaws were still committing evil deeds, first established the [system whereby] official titles were conferred upon [commoners] for their military merits, thereby favoring the soldiers.[24] Today we may follow his precedent and honor this system. Establish an office for the exaltation of the military arts, select [officers by an examination on] *Ssu-ma's Military Arts*,[25] give them the same rank as the Official Erudites, let them give instruction from the Ssu-ma works and administer matters concerning the recruitment, organization, and drilling [of the militia],[26] put them in charge of [conferring] noble ranks and rewards for military merit, and place them directly under the command of the five colonels, *Hsiao-wei*,[27] and ultimately under the control of the Grand

---

[23] The passage explains the second category in *SC* 2.1.

[24] *Wu-kung*, Emperor Wu's edict establishing this practice, was issued in 123 B.C. See *HS* 6:10b–11a, 24B:7; *PC* 24B:8–9a; H. H. Dubs, II, pp. 56–57; Nancy Lee Swann, *Food and Money in Ancient China* (Princeton University Press, 1950), pp. 252–254. The system was generally considered an evil in Former Han times. See also Sakurai Yashio, "Kandai no buko saku ni tsuite," *Tōyō-gakuhō* 26:2 (1939) 78–83. What Hsün Yüeh meant to emphasize here was not the importance of granting noble titles for military merits, but the need to train a civilian militia and eventually revive universal obligatory military service, which had been discontinued under the Later Han. Noble military titles were generally conferred upon commoners rendering or completing their military service; these titles were retained after retirement from active duty. The granting of such titles was thus linked to the development of a commoner-militia.

[25] Now lost; see *Shih-chi* 64:1–3.

[26] Referring to the state-organized spring and winter hunts in ancient times. According to the *Chou-li* 29:7–20, the Minister of War (*Ssu-ma*) used the opportunity of the four-season hunts to organize paramilitary drills for the peasant militia, which culminated in a great game of hunting.

[27] There were eight colonels commanding the imperial army under the Former Han dynasty: (1) *Ch'eng-men hsiao-wei*, Colonel of the Capital Gates;

Commandant, *T'ai-wei*.²⁸ This will not only fulfill [the needs of] the time, but is also in accord with the rites.

By the end of the Chou dynasty, arms had come into frequent [use]. Although social disorder reached a peak during the Ch'in dynasty, still the people had not become completely dissipated [and they retained some of their military training]. Today, the nation has long forgotten [the art of] war. Whenever an emergency occurs or a revolt breaks out all the people are exhausted, on the verge of collapse. "Not to teach the people how to fight [but nonetheless to send them to the battlefield] is to desert them [in war]."²⁹ This is true.

2.7 (4b8) Someone asked: "Of [the systems of] *Chou-mu* (Shepherd of the Province),³⁰ *Tzu'u-shih* (Imperial Inspector),³¹ and *Chien-yü-shih* (Imperial Overseer),³² which is the better one?"

---

(2) *Chung-lü hsiao-wei*, Colonel of the Headquarters [of the Southern Army]; (3) *T'un-ch'i hsiao-wei*, Colonel of the Cavalry; (4) *Pu-ping hsiao-wei*, Colonel of the Infantry; (5) *Yüeh-ch'i hsiao-wei*, Colonel of the Cavalry of the Yüeh People; (6) *Ch'ang-shui hsiao-wei*, Colonel of the Cavalry of the Hu People; (7) *She-sheng hsiao-wei*, Colonel of Archery; and (8) *Hu-pen hsiao-wei*, Colonel of Chariots. See *HS* 19A: 12b–13a. Under the Later Han dynasty, five of these units—(3), (4), (5), (6), and (7)—were placed under the control of the *Pei-chün chung-hou*, Superintendant of the Central Northern Army, and were called the Five Batallions, *Wu-ying*, each commanded by a colonel, *Hsiao-wei*; *HHS*, *chih* 27: 6b–8.

²⁸ Head of the military bureau. The office had been abolished and revived several times during the Han dynasty. See Yü-ch'üan Wang, "An Outline of the Central Government of the Former Han Dynasty," *HJAS* 12 (1949), 150.

²⁹ From the *Confucian Analects*, Legge, I, p. 275. The original passage reads: "To lead an uninstructed people to war is to throw them away," which is slightly different from the *SC* text.

³⁰ *Chou-mu* was first established in 8 B.C. It was abolished in 5 B.C., revived in 1 B.C., abolished again in A.D. 42, and revived a second time in A.D. 188 for the purpose of strengthening the local government, which was threatened by the Yellow Turban Rebellion. Many of these *Chou-mu* became major warlords during the Han-Wei transition period. See *HS* 19: 15; *HHS* 8: 21b, *chih* 28: 1.

³¹ *Tzu'u-shih*, officials dispatched by the central government to make regular inspections of the provinces in compliance with six specific injunctions. This

I said: "These are all governmental systems instituted according to the needs of the times."

Someone asked: "Is not the system of *Chou-mu* now firmly established throughout the empire?"

I said: "In ancient times, the feudal lords each founded their own states; their respective houses held title and authority by heredity. So the administrative power lay therein. [The king established] the worthiest of these feudal lords as their *mu* (Shepherd). [The Shepherds] were in charge of general guidance.³³ They did not do administrative work and were not [responsible for] governing the people. Nowadays the administrators of the commanderies and the magistrates of the districts are appointed and dismissed irregularly, their power is weak and unstable, and the Shepherds of the Provinces wield the greatest political power. The situation is not the same as in ancient times. This is not the way to strengthen the trunk and weaken the branches.³⁴ Furthermore, it is of no practical help in ruling the people. The system of Imperial Overseer is good enough. But of course, if this (the system of Shepherds) were just a measure of expediency to meet the needs of the time, it would be a different [matter]."³⁵

---

system was replaced several times by that of the *Chou-mu* during the Han dynasty. See note 30 above.

³² *Chien-yü-shih* reads *chien-ch'a yü-shih* in the *SC* text. However, since the latter was first established in T'ang times, it must be a corruption of the former which appeared in an identical discourse in *HC* 28:5b–6a. During the Ch'in dynasty, these overseers were permanently stationed in the commanderies.

³³ For the *mu* in the legendary King Shun's time, see Legge, III, pp. 35, 42; in Chou times, see *ibid.*, pp. 525, 530–531; also *Li-chi* 11:16–17a, 19a. *Kang-chi*, formal discipline which may be moral, ideological, or legal.

³⁴ See the important counsel of Chia I (201–169 B.C.) on strengthening the trunk (central government) and weakening the branches (regional power); *HS* 48:10b–12a.

³⁵ The regional levels of administration mediating between the central government and the basic district units (*hsien*) remained unstable until the modern age. In times of unification and peace, a strong central government tended to reduce the power of regional levels of administration by converting permanent provincial offices (such as the *Chou-mu*, Shepherd of the Province

2.8 (5b7) Punishment by mutilation is an ancient practice.[36] Someone said: "Should we revive it?"

I said: "In ancient times the population was large, but now it is very small. To rule a large number of people, one should be stern; to care for a small [population], one should be lenient—this is the Way. The revival of punishment by mutilation is not an [urgent] matter. But there is something we should do: there are people who should have been able to live through punishment [by mutilation], but who are [sentenced to] death instead. For such people, we should revive punishment by mutilation. From times of old, ever since punishment by mutilation was abolished, those who were liable to amputation of their right toe [have instead been sentenced to] death.[37] Only [for such people] should

---

in Han times, or *Sheng*, Province Governor during the Ming and Ch'ing dynasties) into temporary commissions or periodic tours by inspectors; or by dividing a larger administrative unit (such as the *Chün*, commandery in Han times, or the *Tao*, circuit, and *Chou*, prefecture under the Sui and T'ang dynasties) into smaller units, thus increasing their number and decreasing their power. The number of *Chün* was increased from thirty-six under the Ch'in dynasty to over a hundred at the end of the Former Han. The number of *Chou* was increased from thirteen under the Han dynasty to about two hundred under the T'ang, when the status of the *Chou* was reduced to that of the *Chün*. When this happened, especially in times of emergency when the authority of the central government declined, the court would find it necessary to create a new level of provincial government with administrative power over a large territory. This was often done by converting a temporary office of regional inspection into a permanent office, such as the *Chou* in Han times and the *Sheng* in Ming and Ch'ing times. What Hsün Yüeh advocated in this counsel was a compromise measure: establishing a "permanent" office of Overseer for a large provincial area, but making his authority "supervisory" rather than "administrative." This suggestion has merit for present-day China as well.

[36] Punishment by mutilation is mentioned in the *Chou-li* 36:1–2a, 14. This book was an important canon of the Ancient-text School of Confucianism during Han times, but was considered a forgery by the Modern-text School. See also Chang Ts'ang's memorial in *Han-shu* 23:13b–14, Hulsewé, *Remnants*, I, p. 335; Tung-tsu Ch'u, *Law and Society*, p. 174.

[37] Punishment by mutilation was abolished by Emperor Wen of the Former Han dynasty in 167 B.C.; H. H. Dubs, I, p. 255. The circumstances surrounding

punishment by mutilation be revived. This is to give life to [those who would otherwise] die and to allow the population to grow."[38]

2.9 (6a8) [Answer to] a question:

Employment of both moral [education] and statutory punishment is a common practice of the rulers. Sometimes [one will receive priority over] the other because of the needs of different times. The failure to employ moral [education] and statutory punishment is due to an extreme situation.[39] When you first introduce moral education, it should be simple; when you first introduce statutory punishment, it should be lenient.[40] These measures must be carried out gradually. Only after [moral] education has become prevalent and there is no one

---

this move and its grim results are discussed in detail in *HS* 23:12b–14, tr. by Hulsewé, *Remnants*, I, pp. 334–337. He notes: "Hereafter, crimes formerly punishable by amputation of the right foot, moreover, were made equivalent to [those punishable by the] death penalty; [instead of] amputation of the left foot, five hundred strokes of the bastinado were administered. In general, most of the convicted died [from this form of punishment]."

[38] For discussion of the reinstitution of punishment by mutilation during the Han-Wei transition period, see *HHS* 70:19b–13a; *SKC* 10:14a, 11:17b, 13:62–8 (important), 22:4a–5a.

[39] Either the threatening situation or the countervailing policy might be too extreme, *chi*.

[40] *Lüeh*, simple or summary. In traditional China, the legal code was often compared to a fisherman's net. Using a net that was closely woven so as to catch all kinds of fish, irrespective of size, was often considered harsh and oppressive (*mi*), whereas using a net designed so as to catch only big fish was praised as liberal and benevolent (*chien* or *lüeh*). This attitude seems to have had its origin in the Taoist teaching of non-action or non-interference. See A. Waley, *The Way*, pp. 177, 212, 233; Duyvendak, *Tao Te Ching* (London, 1954), pp. 69, 126 (especially note 58), 151. An ideal example of a simple legal code was that issued by Emperor Kao-tsu in 207 B.C. It consisted of only three articles: "He who kills anyone will be put to death; he who wounds anyone, or robs, will be punished according to his offence." See H. H. Dubs, I, p. 58. This principle of minimal legal restraint was later rationalized by the Confucians, who felt that while a legal system was necessary to maintain social order it must be restricted so as to leave a greater field for the operation of moral suasion.

who has not been elevated [to a state of] good conduct, may you enjoin the people in detail.[41] Only after the legal code has been firmly established and there is no one who has not been [admonished] to avoid crime, can you demand strict [enforcement of the law].[42] When it is not yet possible [to enjoin the people] in detail [but you attempt to do so nevertheless], this is called vain instruction. When it is not yet possible to [enforce the law] strictly [but you attempt to do so nevertheless], this is called cruel punishment. Vain instruction harms the reforming influence [of education]; cruel punishment harms the people. A superior man would not follow either course.

To devise [a program of] education [that is so unrealistic] that everyone is likely to avoid it—not considering that the people are as yet incapable of following it—is to incite the people to do evil, and consequently is called harming the reforming influence [of education]. To devise a code of law [that is so rigid] that everyone is likely to commit an offense—not heeding the fact that the people cannot bear it—is to trap the people into the commission of crime, and consequently is called harming the people. When there is no one who has not been elevated [to a state of] good conduct, then it is possible to ask [the people to comply with] the fine points of good behavior; then detailed moral education [can begin]. When there is no one who has not been [admonished] to avoid crime, then it is possible to prevent the people from committing the slightest offence; and only then can statutory punishment be strictly enforced.

2.10 (7a4) Answer to a question: Taking vengeance was a right in ancient times.[43]

Someone said: "Should we leave the people free to take vengeance?"

---

[41] *Tse-pei*, to demand or criticize according to the highest moral standard, or in hair-splitting detail (*pei*, complete).

[42] *Mi*; see note 40.

[43] *Chou-li* 14:11–13a.

I said: "No."

He said: "Then, what should we do about it?"

I said: "There should be both freedom and restriction of action—there should be a chance [for people] to remain alive and [for them] to be killed. These should be systematized according to what is right (*i*) and decided according to law;[44] this is called establishing both right (*i*) and the law."

He asked: "What does this mean?"

I said: "In accordance with the statute on taking vengeance in ancient times, those from whom vengeance for the death of a person's father is sought should be forced to evade [the hand of the avenger by moving] to a different province, one thousand *li* away; for the death of a person's brother, to a different commandery five hundred *li* away; and for the death of a person's paternal uncle or cousin,[45] to a different district one hundred *li* away.[46] He who [succeeds in] taking vengeance upon those who do not avoid him [in this way] should be considered not guilty, whereas he who [succeeds in taking] vengeance upon those who [try to] avoid him [in this way] should be executed. Violating the royal prohibition against [killing] is a crime; taking vengeance is a right, but this right is forfeited by the crime. [Consequently, when vengeance is taken] those who follow the royal ordinance are [considered] obedient, while those who violate the royal ordinance are [considered] disobedient. [The determination of who is not guilty and] should stay alive and [who is guilty and] should be executed ought to be made according to [whether one is considered] obedient or disobedient. [The only exception to the above rule is that] those who go to or remain at their official missions will not be considered as not [trying] to avoid [vengeance]."

---

[44] That is, according to familial morality it is right to take vengeance for the death of a relative by killing the person responsible; on the other hand, according to law, killing a person is a serious crime.

[45] Brothers of one's father and their sons.

[46] *Chou-li* 35:23b.

THE SHEN-CHIEN

2.11 (8a4) Someone inquired about the emoluments of officials.

I said: "The emoluments of the officials of ancient times provided them with sufficient means,[47] whereas the emoluments of the officials of the Han dynasty are set too low.[48] An official's emolument must be appropriate to the position he occupies. If there is one single instance in which [the emolument] is not appropriate, it will not be a [good] system. When officials' emoluments are set too low, private profits are sought. When private profits are sought, honest officials become poor and covetous officials become rich. When the covetous are enriched through private profit-making and the honest are impoverished because emoluments for officials are set too low, the result is confusion and disorder. Therefore the former kings emphasized this matter."

He said: "Should the emoluments of officials be increased?"

I said: "When the wealth of the people has increased significantly, it is proper to increase [the emoluments of officials]."

Someone said: "What about the current emoluments of officials?"

I said: "Times are hard. Officials' emoluments should [be set] according to the means [needed for] subsistence; and the means [needed for] subsistence is related to [the productivity of] the people. These three [factors] are interdependent. If we must [increase officials' emoluments, we should first] rectify corruption in [the dispensation of] emoluments, abolish unnecessary and superfluous offices, adjust with the changing times, be kind and sympathetic to those in low positions, and make reductions or increases [in emoluments] according to the rules; then it will be all right."

2.12 (8b5) Even the feudal lords should not have an absolute [right] to their fiefs. The rich possess under their own names

---

[47] For the Confucian ideal of proper emoluments for officials, see *Mencius*, Legge, II, pp. 373–376; also *Li-chi* 11:1–26.

[48] Lit., too light. For officials' emoluments in Han times, see *Hou Han-shu*, *chih* 28:16b–17. See also Nancy Lee Swann, *Food and Money*, pp. 45–49.

lands in excess of the proper limits, and become wealthier than dukes and marquis—these constitute self-designated [feudal] fiefs. Even the grandees should not have an absolute [right] to their land. But the people may sell or buy land freely on their own; this implies an absolute right to the land.[49]

Someone said: "Should we revive the well-field system?"

I said: "No, absolute ownership of land was not the practice in ancient times, whereas the well-field system is not the [proper] practice for current times."

He said: "Then what should we do about it?"

I said: "Let [the people] cultivate their lands, but they should not have [the right] of ownership; [this should be the policy] until a [more satisfactory] system has evolved."

2.13 (9a7) Someone inquired about the monetary [system].

I said: "The 5-*shu* coins are convenient."[50]

Someone said: "But now they are obsolete. What should we do about it?"

I said: "When [the empire] within the four seas is pacified, the system should be revived."

Someone said: "The coins have been scattered. The region around the capital has been emptied of them. The situation is

---

[49] According to traditional belief, the Chou court claimed that "[of] all under Heaven, none is not the king's land." Hence, no feudal lord under the king could have absolute right to his fief. However, whether the king's claim constituted an ultimate proprietary right to all land or was merely a claim of nominal ruling power over the whole realm is a matter of controversy. In either sense, the government in both traditional and modern China almost always presumes the prerogative of denying absolute proprietary right to its subjects, as evidenced by the idealized well-field system of the Chou dynasty, the land restriction measures (*hsien-t'ien*) of the Han, the land equalization program (*chün-t'ien*) of the T'ang, and the land reforms in Republican and Communist China. For the Confucian utopia of a feudal society based on the well-field system and the problem of land ownership in Han times, see Nancy Lee Swann, *Food and Money*, pp. 66, 113–136, and L. S. Yang, *Studies in Chinese Institutional History* (Harvard University Press, 1963), pp. 92–104.

[50] For a brief survey of the development of the monetary system in Han times including the 5-*shu* coins, see L. S. Yang, *Money and Credit in China: A Short History* (Harvard University Press, 1952), pp. 21–23.

such that these coins must have been hoarded [by people] in distant areas.[51] If we should revive the system, these people would employ the useless coins to trade for our useful goods. This is to exhaust those who are nearby and enrich those who are distant,"

I said: "Such a situation cannot be avoided. What the government desperately [needs] is grain. [Grain, together with] cattle and horses, should be prohibited from leaving the metropolitan area for one hundred *li*. As for other goods, they (the distant people) use their coins to procure something in the east, and then use them in the west. This is to exchange what one has for what one has not, which causes a circulation [of things] around every corner [of the globe]. [In this way] all [the people] within the four seas are [like] a single family. What is there to worry about?"[52]

He said: "The coins have become rare."

I said: "If the coins become rare, the people will care little about them. If after the coins come into circulation again their quantity cannot meet the needs of exchange, then the government should mint [new coins] to make up the deficit."

Someone said: "We should first collect all the coins that have been stored up by the people and transport them to the offices of the Shepherds,[53] who will in turn transport them to the capital. Then [we will have enough coins] to revive their circulation."

I said: "[If you start] a program that is vain and difficult to enforce, the deceitful and defiant will be many, evil and fraudulent practices will emerge, controversies and lawsuits will proliferate, statutory punishments and prosecutions will be intensified, and the sound of sighs, complaints, confusion, and disorder will be heard throughout the world. This is not the way to care for the suffering masses and to bring about peace and prosperity."

[51] See *HHS* 78:34a.

[52] Cf. the liberal economic policy of Hsün-tzu in H. H. Dubs, *The Works of Hsüntze* (London, 1928), pp. 132–133.

[53] 牧 reads 收 in the 1852 and 1917 editions, which follow Lu's *Corrigenda* 1a; but this seems to be misleading.

He said: "Then should we collect and store the coins?"

I said: "No, we should circulate them through market transactions."

Someone said: "Should we remint the 4-*shu* coin?"[54]

I said: "It would be difficult [to do that]."

He said: "Should we then abolish [the use of these coins]?"

I said: "No, the coins are quite convenient for business transactions. The people like to use them, and it would be hard to proscribe them. If we should now begin the difficult program of cutting off the supply of that which is convenient [for the people] and prohibiting the people from doing what they like, it would not bring about prosperity."

He said: "If we should begin to circulate [them] right away, the [number of] coins would not be sufficient.[55] What should we do about this?"

I said: "Whether one honors the coins or proscribes them is dictated [by the demands of the situation].[56] Why should you worry about it?"

2.14 (10b9) The Sage-king should first do [what needs to be done to serve] the people, and then attend to serving the spirits.[57] If the affairs of the people are not yet settled and consequently the sacrificial services in the commanderies[58] are

---

[54] For 4-*shu* coins, see L. S. Yang, *Money and Credit*, p. 22.

[55] Lit., "the money cannot do."

[56] According to Huang's commentary, this means "even if one wants to abolish the coins, it cannot be done," The 5-*shu* coin was reintroduced by Ts'ao Ts'ao in A.D. 208; see L. S. Yang, *Studies*, pp. 191–192.

[57] Cf. Confucius' counsel on serving the spirits; Legge, I, pp. 240–241.

[58] The state cults of the Later Han dynasty are described in the "Treatise of Sacrifices" (*HHS*, *chih* 7–9) as being of three types. First, there were the three most important sacrifices offered by the emperor in person on very exceptional occasions—*kao-t'ien* (reporting the founding of a dynasty to Heaven), *chiao* (Sacrifice to Heaven) at the outskirts of the capital, and *feng-shan* on Mount T'ai (*chih* 7). Second, there were other important sacrifices regularly performed by the emperor (*chih* 8). Third, there were lesser sacrifices performed not by the emperor himself but by his deputies or by local officials (*chih* 9)—offerings to remote and unimportant ancestors of the ruling house who had their temples or tombs in the provinces, regional sacrifices to the legendary *Hou-chi* and his

THE SHEN-CHIEN

neglected, there should be no blame. If we must [perform our services to the spirits] we should first attend to the important ones, making sacrificial offerings accordingly. As for the "distant sacrifices" to the five sacred mountains and four sacred streams,[59] the sacrificial offerings to these spirits have become the established practice of the state in certain districts.[60] If now the commanderies [wish to] perform these sacrificial services, the offerings should be economical. The import of ceremonial rites is to uphold what is fundamental and to show the people that nothing is neglected and nothing is transgressed, thus serving to exemplify the established rituals [of orthodox practice].[61] Provision [for such offerings] should be made in years of good harvest. As for the calamities and strange omens sent down from Heaven [such as the eclipse of] the sun and the moon, these are different from the old [tradition].[62]

2.15　(11b3)　The correspondence between Heaven and man evolves gradually. Between the [first] treading of the hoarfrost and the [arrival of] solid ice,[63] there is a long time that passes.

---

corresponding stars, and other agricultural ceremonies. See especially *HHS, chih* 9:1b–2, 7, 11–14.

[59] *Wu-yüeh*, the five sacred mountains, were: the Eastern Mountain, T'ai-shan, in Shantung; the Southern Mountain, Heng-shan, in Honan; the Western Mountain, Hua-shan, in Shensi; the Northern Mountain, Heng-shan, in Hopei; and the Central Mountain, Sung-shan, in Honan. *Ssu-tu*, the four sacred rivers, were: the *Chiang* or Yangtze River; the *Ho* or Yellow River; the River Huai; and the River Chi (in Shantang). Regular sacrifices to them are mentioned in *HHS, chih* 8:4b–5a. Since these mountains and rivers were distant from the imperial capital, the Han emperor performed the rituals not at these sites but at the capital. Therefore the ceremonies were known as *wang* (lit. "looking toward," i.e. distant sacrifices). See Legge, III, pp. 34–35 (especially notes); *HHS, chih* 8:8b–11a.

[60] *Hsien*, district, in Han times more often signified the emperor (*hsien-kuan*) or his administration. My rendering combines both interpretations.

[61] Cf. *Kuo-yü* 18:1–2.

[62] This sentence is placed by Huang Hsing-tseng in *SC* 2.14, but by context and content it seems to belong to section 2.15.

[63] From *I-ching* 1:23, Legge, *SB*, XVI, p. 60.

Confucius' prayers were not made in a single morning.[64] The eclipse of the sun, [an event] which calls for corresponding action on the part of men, takes place [at irregular intervals]. Sometimes it occurs more frequently than at other times, but it would be unusual if there were two such events in one year. [A passage in] the *Tradition of the Great Plan* reads: "The appearance of the six evil phenomena, if not visible at the imperial court, should not be reported [to the throne]. The respective officials should only mend their own ways."[65] According to the rites of the former kings, [in order to ensure] peace in the royal residence and make possible the removal of such evil phenomena, the Pao-chang and Shih-chin officers should record in writing the occurrence of strange clouds, etc., so as to be prepared [to take appropriate action].[66] The Grand Astrologer should make his reports on this matter [to the throne] without concealing anything. [This matter should] not be neglected.

2.16 (12a9) The son of Heaven, facing south, gives audience to all under Heaven, and at dawn performs his administrative tasks. [The idea for this practice] was probably taken from the *Li* [hexagram in the *Book of Changes*].[67] This is the Way of Heaven. Holding court audience on the first day of each month is an important affair of state. We should rectify the etiquette in connection with it, so as to illuminate this ancient practice.

---

[64] Legge, I, p. 206.

[65] *Liu-li*, six kinds of miasmas produced by the conflict and disorderliness of the Five Elements. According to tradition these miasmas or evil phenomena, which affect the fortune of the Imperial Court, can be avoided or evaded by making sacrificial offerings or instituting administrative reforms. See *HHS, chih* 13:2b-3.

[66] *Pao-chang*, the official in charge of astrological observations mentioned in *Chou-li* 17:15, 26:20-24a. *Shih-chin*, the official in charge of geomancy mentioned in *Chou-li* 17:13b, 25:4-5. For an introductory discussion of these practices, see Joseph Needham, *Science and Civilization in China*, Vol. 2 (Cambridge, 1956), pp. 346-365.

[67] From *I-ching* 9:5a, Legge, *SB*, XVI, p. 426.

THE SHEN-CHIEN

2.17 (12b8) In ancient times there were officials in charge of [the teaching of] the "feminine rites"[68] to the [ladies in the] harem. [These officials] administered the Ordinance for Feminine Education [covering] feminine virtues, feminine speech, [feminine demeanor],[69] and feminine works. They each controlled their subordinates (a group of palace women) and arranged a schedule whereby [they] received favors from the king.[70] This was the rite under the former kings. We should respect the practice so as to give priority to the administration of the harem. The ladies in the harem should peruse the various paintings, read the assorted biographies of [exemplary women of the past], and observe the proper rules of deportment. The Inner Scribe (*nei-shih*) should use her red-tube brush[71] to record the merits and mistakes [of the palace women]; she should examine their behavior and [suggest] demotion or promotion accordingly so as to manifest [His Majesty's] likes and dislikes. When men and women take their proper places outside and inside [the family] and the family is thus rectified, all under Heaven will be set [in good order].[72] When the two elements, *yin* and *yang*,[73] are placed [in proper order], great works will be accomplished. The Way of a superior man should not be neglected for even a day. A superior man must cleave to it even in times of emergency.[74]

[68] The text reads: *yin-yang chih-li*, lit. "rites of the *yin* and the *yang*." The translation here follows Lu's *Corrigenda*, 1b, which reads *yin-li*, "feminine rites." The instruction of the harem in "feminine rites" is mentioned in *Chou-li* 7:12.

[69] *Fu-jung* is missing from the text. Together with the remaining three, *fu-te*, *fu-yen*, and *fu-kung*, they constitute the *ssu-te*, the four virtues of the female in traditional China. Cf. Pan Chao's "Lessons for Women," tr. by Nancy Lee Swann in *Pan Chao: Foremost Woman Scholar of China* (New York, 1932), p. 86.

[70] These two sentences are from *Chou-li* 7:24-25a. *Yü*, an honorific reference to the emperor, later acquired the connotation of referring to the emperor's sexual behavior. "To receive favor from the emperor" thus meant to have intercourse with him.

[71] *T'ung-kuan*, lit. "red-tube"; Legge, IV, p. 69 (especially note to St. 2).

[72] From *I-ching* 4:16, Legge, *SB* XVI, p. 242; R. Wilhelm, *I-ching*, II, p. 215.

[73] *Erh-i*, lit., the two patterns or symbols of *yin* and *yang*, or *ch'ien* and *k'un* (male and female).

[74] See Legge, I, pp. 166-167.

2.18 (13b1) To appoint Official Erudites, to expand the imperial university, and to worship Confucius—these are [proper] rites.

Confucius produced the Classics; thus [they] had a single origin. However, the texts of the Ancient and Modern Schools differ from each other. Both of them claim to be the genuine original version of the Classics.[75] Since the Ancient and Modern [Schools both honored Confucius as their] Former Master, the teachings [of this Former Master] should have been uniform. But the different schools and teachings vary, all of them claiming to embody the genuine ancient [wisdom]. Today, Confucius is remote [from us in time] and no one can approach him with questions. After the old Master died in ancient times, [his teachings] could no longer be heard.[76] [Consequently], who can settle the controversies [over the teachings of the deceased Master]?

When the Ch'in dynasty destroyed traditional learning, [the prescribed] books rotted away inside the [hollow] walls of [private] households and their message was no longer heard in the imperial court or around the country.[77] Down to the rise of the Han dynasty, when efforts were again made to gather and collect the scattered fragments, there did not exist a complete version of the traditional teachings. The written texts were corrupt; there were differences between the [northern and southern] dialects of the Hsia and Ch'u peoples; there were [disparate dates given for the texts, some of which were] recovered earlier than others.[78] Sometimes a hypothesis was tentatively offered by a scholar, and later scholars adopted, elaborated, and ramified it. Consequently one source [branched into] ten streams; [they were as divergent] as Heaven and

---

[75] For the establishment of the Official Erudites and the controversy over the Ancient and Modern Texts of the Confucian canons, see Tjan Tjoe Som, *Po-hu T'ung* (Leiden, 1949), pp. 82-165.

[76] 間 (interval, interruption) reads 聞 (to hear) in the 1852 and 1917 editions. My translation follows Ch'ien P'ei-ming's reading in the 1852 edition, *cha-chi* 2b.

[77] See Derk Bodde, *China's First Unifier* (Leiden, 1938), pp. 80-84.

[78] For a traditional but critical view of the controversy, see Legge's "Prolegomena" to *Shu-ching*, III, pp. 1-34.

THE SHEN-CHIEN

water moving away from each other.[79] The controversies were extremely complex, and the opinions held by [the various scholars] could not all be correct. But if one compares and discusses [the teachings of all the schools], one may come across something noteworthy.

2.19 (14b1) Someone said: "...." The most important Way, [leading man to] the highest virtue, is precise. Yet the written records are abundant. [By exposing oneself to] these works, one becomes versatile and thus attains precision [of thought].[80] There is a saying: "A bird is coming and a net is set up awaiting it. It is one mesh of the net that will catch the bird. But if you now make a net consisting of only one mesh, you will never be able to catch the bird."[81] Although the Way is precise, you will never be able to gain access to it unless by way of versatility. Be versatile in your approach, and precise in your discourse.

2.20 (14b6) The declaration of amnesty is a measure of expedience. Someone said: "Should there be a set rule for it?"

---

[79] *I-ching* 2:5a, Legge, *SB*, XVI, p. 274.

[80] This sentence is corrupt. According to Lu's *Corrigenda*, 1b, the first three sentences belong to the quoted inquiry; 如而 (having access to) in the third sentence should read 如何 (how can one); and Hsün Yüeh's reply begins with the fourth sentence. For the reason for this rendering, see note 65 in *SC* 1.

There is no English equivalent for the Chinese term *yüeh*, translated here as "precision." *Yüeh* originally meant to gird, to restrain (e.g. "the restraint of propriety" in *Lun-yü*, Legge, I, pp. 193 and 220); hence to be moderate and concise (for which the English word "succinct" would be the best translation). However, from "keeping to the most essential" the word derived the meaning of "being the most essential" and "of the greatest importance" in *The Works of Mencius*. See Legge, II, pp. 187–188; also James Ware, tr., *The Sayings of Mencius* (1960), p. 64, and W. A. C. H. Dobson, *Mencius* (Toronto, 1963), p. 85. The precept is related to the notion of *lüeh* (simple and easy) with regard to the legal system, as discussed in *SC* 2.9, note 40. Confucius emphasized "the restraint of propriety" (*op. cit.*), but, unlike Mencius, he did not advocate simplicity as an ideal in itself or as the highest standard of virtue (see *Lun-yü*, Legge, I, pp. 184 and 190). Hsün Yüeh's precept is closer to Confucius' meaning than to Mencius'.

[81] From *Huai-nan tzu* (*SPTK* ed.) 16:12a.

THE SHEN-CHIEN

I said: "There should be no set rules for measures of expedience. Regulations should prescribe the inner justification [of such measures], but not [the specifics of] their execution. 'The *Sun* [Hexagram] should be consulted as the symbol of expedience.'[82] [There should be] regulations justifying [such measures]. A measure of expedience is itself contradictory to what is constant and normal, and has nothing [to do with set rules]."

When [someone] asked for the symbol [from the *Book of Changes*], I said: "The *Wu-wang* [hexagram] shows calamity [happening to one who is free from insincerity];[83] the *Ta-kuo* [hexagram] shows that there will be evil in excess.[84] These are the pertinent symbols."

[Amnesty] should be granted only as a last resort; there should be a prohibition on [invoking it] too frequently.

Someone said: "Should we then do away with [amnesty] altogether?"

I said: "[When a measure is] deemed expedient or convenient [in an emergency], you cannot do away with it altogether."

2.21 (15a7) The practice of *Shang-chu* is not an ancient [rite].[85] According to the *Canon of T'ao-t'ang*,[86] [the Sage-king] issued orders sending his two daughters [to the north of Kuei to be married into the family of Yü].[87] According to the instructions of Ti-i [of the Shang dynasty],[88] his younger sister was sent to marry a vassal so as to ensure the happiness and great good fortune [of the state]. According to the rite of the Great Chou, the king also sent his royal princess to marry

---

[82] *I-ching* 8:18b, Legge, *SB*, xvi, p. 398; R. Wilhelm, *I-ching*, i, p. 371.
[83] *I-ching* 3:23b, Legge, *SB*, xvi, p. 110.
[84] *I-ching* 3:32b, Legge, *SB*, xvi, pp. 117, 236.
[85] See note 5. For further information see *Hsün Yüeh*, pp. 153–154.
[86] The legendary King Yao.
[87] Legge, iv, p. 27. The family of Yü, to which the legendary King Shun, who succeeded King Yao, belonged.
[88] *I-ching* 2:13b, Legge, *SB*, xvi, p. 82; R. Wilhelm, *I-ching*, i, p. 53 and note. Ti-i was the 29th ruler of the Shang dynasty.

the [Marquis of] Ch'i.[89] For the female to preside over the male is to contradict the [will of] Heaven; for the wife to overpower the husband is to contradict the [will of] Man. To contradict the [will of] Heaven is inauspicious; to contradict the [will of] Man is unjust.

2.22 (15b5) In ancient times, the son of Heaven and the feudal lords had to report to their [ancestral] temples whenever someting [important] occurred. At their courts there were two Scribes (*Shih*): the Left Scribe (*Tso-shih*) recorded their speech, and the Right Scribe (*Yu-shih*) recorded their actions.[90] [The records of] their actions became the *Spring-and-Autumn Annals* (*Ch'un-ch'iu*); the records of their speech became the *Historical Documents* (*Shang-shu*). The ruler's every action and its merit or lack of merit, its success or failure, had to be written down. [From the emperor] down to the common people, whoever achieved outstanding merit[91] was also noted. Those who wished to make themselves well known were not allowed to do so; those who desired to conceal [themselves] only succeeded in having their names exposed. The meritorious or blamable deeds committed in a single morning would become a record of glory or disgrace for thousands of years. This would encourage those who were good and frighten those who were evil. Therefore the former kings stressed [this practice] and used it as a means of supporting [their program of] law enforcement and [moral] education.

What we should do now is to oblige the various officials to keep daily records of their respective activities.[92] At the end of

[89] The "Little Preface" to the ode *Ho p'i-nung*, Legge, IV, "Prolegomena," p. 41.

[90] The text is corrupt. The *SC* quotation in *CSCY* 46:6b reads: "The Right Scribe recorded events (*shih*); the Left Scribe recorded speech." Cf. *HS* 30:8a. See Han Yü-shan, *Elements of Chinese Historiography* (Hollywood, California, 1955), p. 2.

[91] The text 等各有異 "[they] were of different ranks," reads 苟有茂異, "whoever achieved distinctive merits," in *HHS* 62:14a and *CSCY* 46:6b.

[92] The character 日 (daily) reads 方 (directions, aspects) in *CSCY* 46:6b. My translation combines these two versions.

a year,[93] these records should be sent to the Master-of-Writings (*Shang-shu*). The Scribes, in accordance with the established precedents of their respective offices,[94] should be in charge of this matter. They should not write things which seditiously depart from the accepted standards. [Other than such matters], any good or evil behavior should be recorded; all speech or actions that may serve as models or precedents should be recorded; all meritorious deeds should be recorded; all military operations affecting the masses should be recorded; the coming of the Four Barbarians to the imperial court with their tributes should be recorded; the appointment or establishment of the empress, the royal concubines, or the heir-apparent should be recorded; the appointment or dismissal of imperial princesses[95] and high-ranking officials should be recorded; good and bad luck and any licentious or unruly behavior [of the ruling house] should be recorded; favorable omens and disastrous portents should be recorded.

Our former emperors set the precedent of compiling a Record of the [Emperor's] Daily Life, *ch'i-chü-chu*.[96] Every detail of his daily activities, [from] his actions to his repose, had to be noted down. This practice should be revived. The Inner Scribe, *Nei-shih*, should take charge of the matter and record the events of the inner [palace].

---

[93] 其 reads 歲 in *CSCY* 46:6b; my translation follows the latter version.

[94] Collation from *HHS* 62:14b and *CSCY* 46:6b.

[95] *Kung-chu* (princess) may also be a corruption of *kung-wang* (duke and prince).

[96] *Ch'i-chü* (lit. rising and resting) refers to the daily life of the emperor. The compilation of such a record in Former Han times is mentioned in *Pao-p'u tzu*; see *CC* 10A:9b. In *HS* 30:7b–8a, a *Chu-chi* of 190 *chüan* is recorded. A *Ch'i-chü chu* of Hsien-tsung (Emperor Ming, r. 58–75) compiled by his Empress née Ma was mentioned in *HHS* 10A:15. The *Ch'i-chü chu* of Emperor Hsien, which seems to have been compiled at Hsün Yüeh's request, is extensively quoted in *HHS* 9:1b, 12a; 62:18a; and *pass*. See also Charles S. Gardner, *Chinese Traditional Historiography* (Harvard University Press, 1938; 2nd printing, 1961), pp. 88–90; Hans Biehenstein, "The Restoration of the Han Dynasty," *Bulletin of the Museum of Far Eastern Antiquity* 26 (1954), 21–23.

Shen-chien 3
Common Superstitions
(*Su-hsien*)

3.1 (1a4) Someone inquired about [divination using] tortoise shells and yarrow sticks.[1]

I said: "A virtuous [person] would benefit by it; others would only be injured."

He said: "What is the meaning of this?"

I answered: "When an omen signifies good fortune and one can take advantage of it [because of one's virtue], or when it signifies bad luck and one can be saved [from disaster because of one's virtue], this is called benefit. When an omen signifies good fortune and one becomes complacent, or when it signifies bad luck and one becomes discouraged and negligent, this is called injury."[2]

3.2 (1a7) Someone inquired: "What about the various superstitions of the time?"[3]

---

[1] This refers to two methods of divination: the "oracle bone" divination of the Shang dynasty and the "number-mysticism" of milfoil (yarrow) counting elaborated in the *I-ching*, which probably originated during the Chou dynasty. Here it may mean divination in general.

[2] In this and the following discourses, the emphasis is on the inner virtue of man. Although Hsün Yüeh did not deny the influence of nature or of the cosmos on men's fortunes, he argued that in cases where their influence on various people was the same, it was the difference between the inner worth of the individuals that made a difference in their fates. Thus he thought that the workings of nature and the cosmos were beyond human control and that a man's virtue might not influence them per se; but on the other hand, a man's inner goodness might result in a better orientation of his existence in the world. Hence the repeated references to the outer sphere (nature and the cosmos in time and space—form) and man's inner strength (virtue, strength of character, essence of personality—substance).

[3] According to Lu's *Corrigenda* 2a, "time" should read "day and time (hours)."

THE SHEN-CHIEN

I said: "These [superstitions] concern the cosmic numbers[4] of Heaven and Earth. They have no effect on the good or bad fortunes of men. The east signifies growth, but no fewer people die [in that quarter than elsewhere]; the west signifies destruction, but no fewer people are born [in that quarter than elsewhere]; the south signifies fire, but those who live [in that quarter] are not burnt; the north signifies water, but those who walk there are not drowned.[5] The dawn of the *chia-tzu* day simultaneously saw the extinction of the Yin dynasty and the rise of the Chou dynasty;[6] Hsien-yang is both the place where the Ch'in dynasty was destroyed and the Han dynasty prospered."[7]

3.3 (1b6) Someone said that the particular positions of the Five [planets] and the Three [luminaries in the celestial sphere] signaled [the rise of] the Chou dynasty,[8] and the

---

This refers to the Han belief in astrology—that the outcome of man's action is determined by the position and movement of the sun, the moon, and the stars during a particular day or hour.

[4] *Shu*, numbers; see note 47 in *SC* 1.

[5] The equations are based on the Five Elements theory, see J. Needham, *Science*, Vol. 2, pp. 232–273.

[6] According to the "*Wu-ch'eng*" (Successful Completion of the War) chapter in the *Shu-ching*, the Chou army defeated the Shang forces at Mu on the *chia-tzu* day and thereby overthrew the Shang dynasty. See Legge, III, pp. 314–315.

[7] Hsien-yang, near present Sian, was the capital of the Ch'in dynasty. It was destroyed in 206 B.C., and near its site in 202 B.C. was erected Ch'ang-an, the new capital of the Former Han dynasty.

[8] "Five" refers to the five stars or constellations in the Five Elements cosmology: (1) Jupiter–East–Wood–*Sui-hsing*–Blue Dragon, (2) Mercury–North–Water–*Ch'en-hsing*–Black Tortoise, (3) Venus–West–Metal–*T'ai-po*–White Tiger, (4) Mars–South–Fire–*Ying-huo*–Red Bird, and (5) Saturn–Center–Earth–*Chen-hsing*. "Three" refers to the sun, the moon, and the star (polar?). The positions of these celestial bodies were believed to have a strong influence not only on the natural environment but also on men's fortunes, especially the fortune of the ruling dynasty. Therefore each new dynasty would devise a new calendar to accord with the celestial cycles. See *HS* 21A:14b–15a; *HHS* 30B:8a–17b. The dates and the celestial phenomena relating to the founding of the Chou dynasty are discussed in *HS* 21B:20–21a; also *PC* 21B:

THE SHEN-CHIEN

union of the "Dragon-tail" represented an auspicious omen for the Chin state.[9]

I said: "[The universe may be likened to mankind's abode.][10] The furnishings of an official mansion suit the status of a wealthy and honorable [official]; a commoner occupying it would find it inappropriate to his rank. A jail consists of locked chambers and chains, but only criminals are committed to them; an innocent person may enter [and leave] without such suffering."

He said: "Then, should the [astrological calculation of the calendar based on the position and movements of] the sun and the moon be abolished?"

I said: "No, the basic method of calendric calculation[11] was used by the former Sage-kings. Man follows Heaven and Earth. Therefore, his [periods of] activity and quiescence should coincide with theirs; that is, [his cycles of activity should] accord with the principles of *yin* and *yang*, with [the

---

51b–60a. For a modern study see J. Needham, *Science and Civilization in China*, Vol. 3 (Cambridge University Press, 1959), pp. 194–201, 232–253, 390–406.

[9] This sentence alludes to Chin's conquest of Kuo in 654 B.C., recorded in the *Tso-chuan*; see Legge, V, pp. 144–416. 龍虎 (Dragon-tiger) reads 龍尾 (Dragon-tail) in the *Tso-chuan*. These stars are mentioned in note 8: Dragon (1)—Jupiter; Tiger (3)—Venus; and Tail (*Wei*)—a component of the Dragon constellation. The *Tso-chuan* version reads, "The dragon's tail lies hidden in the conjunction of the sun and the moon." See also J. Needham, *Science*, Vol. 3, p. 409.

[10] In the West, life is often likened to a pilgrimage. The Taoist in Wei and Chin times often compared man's existence in the universe to a traveller's stay at an inn. Thus it is related that the eccentric Liu Ling (fl. early third century) once lay naked in his house. When someone reproached him, he replied: "Heaven-and-Earth is my abode; this house is but my clothes." *Shih-shuo hsin-yü* (*SPPY* ed.) 3A:29a; also Fung Yu-lan, *A Short History of Chinese Philosophy*, p. 235. In this passage, Hsün Yüeh used the metaphor to stress the importance of the inner virtue of a man in contrast to outside influences; see note 2.

[11] *Yüan-ch'en* read *yüan-jih yüan ch'en* 元日元辰 in Lu's *Corrigenda* 2a. According to Huang's commentary, *yüan-ch'en* refers to the special positions of the sun and moon at the completion of a celestial cycle, when a new calendric cycle should begin.

position and movement of] the sun and the stars, and with the calendric and other numerological rules [derived from them]. Inwardly, there should be harmony of substance; outwardly, there should be harmony of form.[12] This harmony of form and substance is [ordered by] reason.[13] Omens that make one happy and are auspicious are signs of nature's response [to such a state of concord]. Confucius refused to drink at the Brigand-spring[14] and Mo Ti refused to enter [the city named "Morning-singing,"] Chao-ko;[15] they disliked the names [of these places] and acted according to their conscience. [A man who] lacks substance (strength of character) merely seeks good fortune by superstitious means. This is difficult to accomplish."

3.4 (2b8) Someone said: "How should the spirits be approached?"

[I said:][16] "Only devotion will incur a natural response. Therefore [we must] support [our prayers] with mental devotion, manifest them through sacrificial offerings of animals,

---

[12] *Shih* (substance; the inner worth, moral virtue, and strength of character in man) versus *wen* (outer display, brilliance of appearance, form and formality indicative of the effect of culture). Here Hsün Yüeh compromised his position on the inner worth of man as the decisive factor, admitting that man's action should be in accord with the outside world (nature, the cosmos). But he emphasized that the accord must come from within, that inner harmony of character should substantiate the outer harmony of form.

[13] 理 *li* means (1) to order and administer (by human effort), or (2) the cosmic and metaphysical principle, or Reason, which prescribes human reason and order. The term was frequently used in the first sense in pre-Han and Han texts, but the second meaning became predominant in Neo-Confucian writings. My translation is a compromise between the two interpretations, and is justified by Hsün Yüeh's use of the term in *SC* 5.12 and 15, where it means "reason."

[14] *Shui-ching chu* (Wu-ying tien chü-chen pan ed.) 25:27b–28a; *Wen-hsüan* (*SPTK* ed.) 28:1b; *HHS* 41:15a.

[15] *Huai-nan tzu* 16:11b. *Chao-ko* (lit., Morning-singing), capital of the Shang dynasty in its later days. Mo Ti advocated frugality and hard work and condemned the Confucian rituals and music as wasteful; hence he abhorred the name of the city, which conveyed a sense of decadence.

[16] Translation follows Lu's *Corrigenda* 2a, which differs slightly from the text of the *SPTK* ed. and the quotation in *PTSC* 90:8a.

jade, and fine clothes, and then [seek to] communicate with the spirits through sincere entreaty. What is meant by the ritual [of sacrifice] is not just the offering of jade or fine clothes; what is meant by praying is not merely [the offering of] wine or food. [Sacrifices] not in accord with the rites may result in transgression; improper prayers will not be answered."

3.5 (3a4) Someone inquired about whether praying [has any effect].

I said: "[An effect] is possible on the basis of the correspondence between *ch'i* (the ether) and *wu* (material things).[17] But it is futile [to pray] where [matters of human] nature and destiny and [the laws of] nature are concerned."

3.6 (3a6) Someone inquired about whether it is possible to avoid illness and danger.

I said: "What is the cause of illness and danger? It lies either in the body or in the spirit. One cannot avoid one's body, nor can one evade the spirit. That which can be avoided is not the body, neither is that which can be evaded the spirit. One always carries one's body, and follows Heaven. No matter how many thousand *li* a man travels, he can never escape the company of these two. Think of a child covering his eyes under the armpit of a big man—could that possibly be called an escape?"

3.7 (3b1) Someone inquired about physiognomy.

I said: "There might be [some value in] this practice. [Man's] spiritual essence and physical appearance[18] are by nature interrelated. They are affected in complex ways by his conduct and by [the changing] times. [The different factors involved] complement and constrain each other, making the

[17] *Ch'i*, air, breath, "ether," which constitutes the essence of the *wu* of all beings. See *SC* 5.16 and notes.

[18] *Shen-ch'i*, spirit (mind) and essence (of matter); *hsing-yung*, appearance or shape; see *SC* 5.16 and notes.

number [of their possible combinations] and variations[19] exceedingly great. Basically, they (the varieties of man's essence and physical appearance) can be divided into [three types]: upper, middle, and lower."[20]

3.8 (3b5) Someone inquired about the art of [becoming] immortal.[21]

I said: "How pretentious [are those who talk about the art of becoming immortal]! The less one is concerned with it [the better]. The Sage does not study it.[22] This is not because he despises life. [It is because] the beginning and the end [of all things] are [the result of] cosmic cycles.[23] Long or short life is determined by fate.[24] The workings of nature and fate[25] are not affected by human effort."

He said: "Are there immortals?"

I said: "The pygmies and the cassia jungles[26] are products of strange environments. [Likewise], even if there were immortals, they would merely be a different kind of being."

---

[19] *Shu*, numbers, lot, fortune, fate; *pien*, changes and variations. See notes 47 and 72 in *SC* 1.

[20] See *SC* 5.15.

[21] *Shen-hsien* (*Shen*, spirit; *hsien*, those who have transcended physical existence and become immortal). Cf. Wolfgang Bauer, *China and the Search for Happiness* (1976), pp. 100–108.

[22] Legge, I, pp. 150 (?), 201, 240–241.

[23] *Yün*, cyclical movement, later developed the connotation of the workings of fate.

[24] *Shu*, see note 19.

[25] *Yün-shu* later became an established term for fate in colloquial Chinese.

[26] *Chiao-yao*, pygmies, described by Confucius as 3 feet (Chinese measure) in height; *Shih-chi* 47:7a, tr. by Edouard Chavannes, *Les Memoires Histoires de Se-ma Ts'ien* (Paris, 1895–1905), v, pp. 314–315 and note. They were described by Lieh-tzu as being 1 foot, 5 inches; *Lieh-tzu* (*Han-Wei ts'ung-shu*, Fourth Series, 1880) 5:4b, tr. by A. C. Graham, *The Book of Lieh-tzu* (London, 1960), p. 98. In the same section in *Lieh-tzu* many other extraordinary beings are mentioned; Graham, pp. 92–117. *Kuei-mang*, cassia jungle, forest of fragrant trees; see the couplet 徑乎桂 (*kuei*) 林之中過乎泱莽 (*mang*) 之野 in *Shih-chi* 117:13a. See also *Shan-hai ching* (*SPTK* ed.), II, 38a.

3.9 (4a4) Someone inquired about whether there were [in fact] people [who lived to be] several hundred years old.

I said: "Wu Huo was renowned for his [extraordinary] physical strength;[27] Chiang Hai, for his swift-footedness;[28] [Meng] Pen and [Hsia] Yü, for their prowess;[29] Confucius, for his wisdom; and Ancestor P'eng, for his longevity.[30] That there are beings of unusual qualities cannot be denied."

3.10 (4a9) Someone inquired about whether longevity was predestined by the Way and could not be attained by human effort.

I said: "Those who attain longevity are those who benefit by the Way; and those who benefit by the Way are those who are by nature endowed with longevity. If it is not endowed by nature, no one can attain it through training. Nevertheless, he who studies to be a Sage will at least be able to develop his nature (*hsing*)[31] to the fullest extent; and [he who attempts to attain] longevity in accordance with the Way will at least be able to develop to the fullest extent what is destined (*ming*) by nature."[32]

---

[27] A man of exceptional prowess at the court of King Wu of Ch'in (309-306 B.C.). See Legge, II, p. 425 and note; *Shih-chi* 79:6a.

[28] According to Huang's commentary, Chiang Hai should read Shu Hai 豎亥 who, according to a legend in *Shan-hai ching*, II, 46b-47a, was commissioned by King Yü of Hsia to measure the earth by walking from its eastern extremity to its western extremity. According to Lu's *Corrigenda* 2a, Chiang should read Chi 忌, alluding to Ch'ing-Chi, son of King Liao of Wu, who could catch the flying bird and challenge the fierce beast; *Wu-Yüeh ch'un-ch'iu* (*Han-Wei ts'ung-shu*, Second Series) 2:5b, 8-9a. However, *Shih-chi* 79:6a mentions Ch'eng Ching 成荊, an ancient hero who was also called Chiang 羌.

[29] Meng Pen, a warrior who could pull the horn off a bull; Legge, II, pp. 185-196 and note. Hsia Yü, a man who could lift a weight of 30,000 catties; *Shih-chi* 79:6a.

[30] P'eng-tsu, legendary founder of the state of P'eng who, it is said, lived for more eight hundred years during the Hsia and Shang dynasties. See *Lieh-hsien chuan* (*Ts'ung-shu chi-ch'eng* ed.), I, pp. 14-15; *Le Lie-sien tchouan*, Fr. tr. by Max Kaltenmark (Peking 1953), pp. 82-84.

[31] *Hsing* reads 生 *sheng* (life) in *I-lin* 5:68. The two words are interrelated; see Wolfgang Bauer, *op. cit.*, pp. 36-38.

[32] *Ming*, destiny. See Fu Ssu-nien, *Hsing-ming ku-hsün pien-cheng* (Shang-wu

3.11 (4b4) Someone said: "Is it true that there are some human beings who [are able to] transform themselves and become immortal?"[33]

I said: "I have never heard of this before. Even if [such a metamorphosis] did occur, [the result would be] a monster, not an immortal. There have been cases of a man being changed into a woman[34] and a dead person regaining his life.[35] But is this the nature of human beings? It has nothing to do with the cosmic numbers."[36]

3.12 (5a1) Someone asked: "Is it possible to nourish one's [inner] nature?"

I said: "To nourish his [inner] nature, a man should hold fast to [the precepts of] the golden mean and harmony, and safeguard [the vital energies of] life."

To treasure (*ai*) the emotional (*ch'in*), the moral (*te*), the physical (*li*), and the intellectual (*shen*) energies[37] is called being miserly.[38] When this is incorrectly practiced, there is no outlet [for one's vital energies]; and when this [practice] is taken to the extreme, tranquility (balance) will be disturbed. Therefore, a superior man should temper the expenditure of

---

yin-shu kuan, 1940) 2:36–62 and *pass.*; also Wolfgang Bauer, *op. cit.*

[33] Cf. Wolfgang Bauer, *op. cit.*, pp. 100–108.

[34] *HHS* 9:13a; the incident was recorded in A.D. 202, three years before the completion of the *SC*.

[35] *HHS* 9:12b; the incident was recorded in A.D. 199, six years before the completion of the *SC*.

[36] *Ch'i-shu*, essence-numbers, later became a standard term for the "common fate" or destiny of a state or an age.

[37] The meaning of this sentence is not clear. *Ai*, to prize, to care for; *ch'in*, the intimate, the self, the emotional; *te*, virtue, power, function; *li*, physical and bodily strength; *shen*, spirit, intellect.

[38] *Se*, miser, to spare; see *Lao-tzu* 59. Rendered by Legge as "moderation," *SB* XXXIX, p. 102. Arthur Waley's translation, "laid up a store," in *The Way and Its Power*, p. 213, accords with the definition given in *Shuo-wen Chieh-tzu ku-lin*, 2297–2299. Ho-shang Kung's commentary on this line from *Lao-tzu* expressed a similar concept, that of saving the "vital energies"; *Lao-tzu ku-chu*, II, 28b. For the technique of "storing the energies," see note 43 below.

his vital energies so that no obstruction, congestion, or exhaustion[39] will upset the hundred departments [of his body] and cause illness.

In joy or in anger, in sorrow or in happiness, in [active] thought or in [reflective] contemplation, one should maintain the golden mean.[40] This is to nurture the spirit. In cold or warm weather, during the new or full moon, and throughout the changing seasons one should always maintain the golden mean. This is to nurture the body.[41]

He who is well-versed in regulating the vital energies is like King Yü regulating the water-works.[42] But he [who practices the art of] guiding the inner breath, storing the vital energies, and protecting the vital organs [of the body] through the [technique of] calling forth inner visions[43] will tend to overdo it, and consequently violate the golden mean. These techniques may be used to cure some diseases, but they are not [the same as] the Sage's art of nurturing his [inner] nature. Holding the [breath] may be [practiced] to expand it; storing up [the energies] may be [practiced to make up for] a deficiency; the calling forth of inner [visions] may be [practiced] to complement external [impressions]. When the energies should be [freely] flowing, but are [consciously] impeded; [when] the body should be [left alone in a state of] harmony, but is [consciously] regulated by exercise; [when] the spirit should be at peace, but is [consciously] exerted through [visionary] practices, the result will be a loss of harmony. Those who know well how to nurture their [inner] nature [realize there is] no other permanent solution than to maintain [an inner] harmony.

---

[39] The sentence is from *Tso-chuan*, Legge, v, p. 580. A similar precept is given in *Kuo-yü* 18:4.

[40] Legge, I, p. 384.

[41] I.e., by channeling rather than obstructing the flow of the body's energies. See Legge, v, pp. 580–581.

[42] Legge, III, pp. 92–150.

[43] For the Han and post-Han Taoist techniques of longevity, see Henri Maspero, "Le Taoisme," *Mélanges posthumes sur les religions et l'histoire de la Chine* (Paris, 1950), II, pp. 81–116. A summary account of the Taoist breathing and sexual techniques for longevity is given by J. Needham in *Science*, Vol. 2, pp. 143–152.

THE SHEN-CHIEN

The region within a two-inch area of the navel is called "the pass."[44] "The pass" is the place [where one] takes in breath and passes it on to the four parts of the body. Therefore, he who [wants to] expand his breath takes it in through "the pass." He who takes shorter breaths uses the higher parts of the body; as a result, his pulse becomes quicker and his mind agitated. By taking in breath through the shoulder,[45] a person's breath flows more freely and he is better able to concentrate. By breathing through "the pass," the breath flows most easily. Therefore, the practitioner of Taoism always directs his breath through "the pass." This is called the vital art.

The *yang* element is life-nourishing; the *yin*, reductive and destructive. The essence of a cordial and genial person is *yang*. Consequently, he who wants to nourish his nature encourages that which is *yang*, and suppresses that which is *yin*. But when the *yang* element is excessive, one becomes overly aggressive. Similarly, when the *yin* element is excessive, one becomes stagnant. Being overly aggressive leads to remorse; being stagnant leads to calamity.

The myriad things cannot generate [the energy of] spring; they must wait for the spring to bring them to life. But man is different—he need only preserve the spring [within himself].[46]

Medicine is used to cure illness. When one is not ill, one need not take medicine. A man should not consume meat in amounts disproportionate to rice.[47] How much more detrimental it is to

---

[44] *Kuan*, gate, stronghold.

[45] The character *chien* 肩 seems to be a corruption. The breathing exercise progresses through the following stages: breathing through the throat, through the chest (or shoulders), through the upper belly, through the lower belly ("pass"), and finally through the heels (by meditation).

[46] *Ch'un*, spring, the principle and capacity of growth, has a strong sexual connotation. Thus "the storing up of the vital energies" and " the preservation of the spring" come very close to the sexual technique of conserving the seminal essence (*ching*) and the divine element (*shen*), as discussed by J. Needham. See note 43.

[47] Legge, I, p. 232; 氣 air, reads 餼 rice; see *Lun-yü ku-chu chi-chien* (Ch'iang-su shu-chü 1881 ed.) 5:64b–65a. Cf. the manual on "Dietary Regimen and Breathing Exercise" recently discovered from the Former Han tomb at Ma-wang tui, *Wen-wu*, 1975, No. 6, pp. 1–19.

take [too much] medicine! A cold produces a fever; and a fever causes congestion and indigestion. Use of a reductive medicine[48] [in this case] is appropriate and will cause no harm. But when the vital energies are in balance, taking medicine will upset [the balance]. The use of acupuncture and cauterization should follow the same principle. Therefore, those who [seek to] nurture their [inner] nature should not take a great deal of [medicine]. The precept to remember is temperance.

3.13 (6b8) Someone inquired about why the virtuous live long.[49]

I said: "A virtuous person neither inwardly injures his nature nor outwardly damages things; he acts neither against Heaven above nor against men below; he behaves properly and resides within a [harmonious] center [of being]; he maintains harmony between body and spirit. Consequently he incurs no evil omens and accumulates all kinds of blessings. This is the art of longevity."

Someone said: "How about Yen [Hui] and Jan [Po-niu]?"[50]

I said: "This is fate (*ming*):[51] the wheat plant cannot outlast the summer and a flower cannot outlive the spring, no matter how harmonious the elements are. Although [the lives of Yen and Jan] were short, [nevertheless they lived as] long [as they could] within [their destined life-spans]."

3.14 (7a8) Someone inquired about the alchemist's [formula for producing] gold and silver.[52]

---

[48] *Yin-yao*, medicine which dilutes and cleanses the congested humours of the body, the way a fire-extinguisher quenches a fire (*yang*).

[49] Legge, I, p. 192.

[50] Two virtuous disciples of Confucius, one short-lived, the other affected by chronic disease; Legge, I, pp. 185, 188.

[51] See note 32.

[52] *Huang-pai* (gold and silver) refers to alchemical experiments to produce gold from cinnabar; *Shih-chi* 12:3b–4a, 9; *HS* 36:6b. The topic is discussed in detail by Ko Hung in *Pao-p'u tzu* (*SPTK* ed.) 4:1–21. See also H. Maspero, *Mélanges posthumes*, II, pp. 96–98.

I said: "Fu I's comment on such formulas is correct.[53] It is possible to bake a piece of clay and produce a tile, but it is impossible to burn a tile and change it into copper. [Some people] make inferences from [the observation of] natural [changes and suggest that it is possible to bring about changes that are] unnatural or impossible. How deceptive![54] It is impossible[55] to produce a horse or an ox by matching[56] the muscles of dogs and sheep, is it not?"

3.15 (7b7) Although people have asserted that the *wei* books were written by Confucius,[57] the late *Ssu-k'ung*, Shuang,[58] an uncle of your humble servant Yüeh, investigated [the matter] and found this to be untrue. The books were probably written by Chung Chang and his followers[59] before the restoration of the Han.

Someone said: "There might be a mixture [of the genuine writings of Confucius and later forgeries in these books]."

I said: "[By 'mixture' do you mean] that some later authors inserted their writings into Confucius', or that they inserted some of Confucius' writings into theirs? [I think] these works are [by later authors] who inserted some of Confucius' writings

---

[53] Fu I (ca. A.D. 47–92); biography in *HHS* 80A:13b–16a.

[54] See Ko Hung's vigorous defense of such alchemical experiments in *Pao-p'u tzu* 2:2–4a, tr. by J. Needham, *Science*, Vol. 2, pp. 437–439.

[55] *Chi*, near, likely, hopeful. Legge, v, pp. 845–847; *Shih-chi* 39:36b.

[56] *Ti*, contrast, compare, match.

[57] *Wei*, the apocryphal books allegedly containing the esoteric teachings of Confucius, in contrast to the Five Classics which were described as his exoteric teachings. Cf. Tjan Tjoe Som, *Po Hu T'ung*, pp. 95–128. See also *SC* I, note 1.

[58] Hsün Shuang (A.D. 128–190), biography in *HHS* 62 (*lieh-chuan* 52), wrote a *Pien-ch'an* (Criticism of the Prognostic *ch'an* Works), now lost. See Introduction, pp. 33–34; also Chi-yun Chen, "A Confucian Magnate's Idea of Political Violence: Hsün Shuang's Interpretation of the Book of Changes," *T'oung-pao* 54 (1968), 73–115.

[59] The name Chung Chang is not in the histories of the Former and Later Han dynasties. The only person with the surname Chung mentioned in these works is Chung Chün, an eloquent man of letters who had discussed the *Ch'an-Wei* omens with Emperor Wu, and who wrote a book in eight *p'ien* (chapters);

into their own. Therefore, we may say that the eighty-one chapters of the *wei* books[60] are not the writings of Confucius."

Someone said: "Should we burn them?"

I said: "Although we cannot say that these books are the works of Confucius, there are still things in them worth learning. Why should we burn them?"

3.16 (8b4) A ruler should not accept unfounded statements, heed pompous counsel, adopt flowery names, nor start unjustified projects. Advice must be practical; strategy must be based on precedent; names must tally with substance; government programs must produce good results.

HS 30:14a, 64B:4b–8. Another important term recorded in the two dynastic histories is *chung-shih* (beginning-end), the theory of "cosmic cycles" expounded by the *Yin-yang* School and later affiliated with the *wei* books. See *HS* 30:16b; *HHS* 1B:25b, 14:10a, 28B:5b, 35:13a, 40B:4a. Also Tjan Tjoe Som, *Po Hu T'ung*, pp. 100–102.

[60] *Ho-t'u* in 9 *p'ien*, *Lo-shu* in 6 *p'ien*, other similar works in 30 *p'ien*, and the seven *wei* of the canons in 36 *p'ien*. For a complete list, see Tjan Tjoe Som, *Po Hu T'ung*, pp. 102–104.

Shen-chien 4
Miscellaneous Dialogues
(Tsa-yen), I

4.1 (1a4) Someone asked: "Why should a superior man earnestly encourage learning?"

I said: "Those who are born with knowledge are few; those who know by learning are many.[1] [The difference between] the multitude of commoners and the gentlemen of leisure,[2] or between enlightened government and dark anarchy[3] depends upon the success or failure of learning. Is it not appropriate that one should earnestly encourage learning?"

4.2 (1a8) A superior man should reflect upon three things: the past, [other people], and [what he sees in] the mirror.[4] [Looking at] the past, he observes the lessons [of history];[5] [looking at other] people, he learns from those who are virtuous; [looking] in the mirror, he sees a clear reflection [of himself]. The decline of the Hsia and Shang dynasties was due to their failure to heed the lessons of King Yü and King T'ang;[6] the ruin of the Chou and Ch'in dynasties was due to the failure to

---

[1] *Kua* (few) in the second clause is a corruption. My translation follows Lu's *Corrigenda* 2a. For Confucius' discussion of inborn knowledge and learning, see Legge, I, p. 201.

[2] *Yu-yu* and *i-i* both mean multitudinous and easy. See *Shih-chi* 47:18b, tr. by Edouard Chavannes, v, p. 363 and note; *HHS* 43:10b; Legge, IV, pp. 290, 500, 169 (note to line 2), 51–52 (note).

[3] *Wen-wen*, dirty, unenlightened; *Shih-chi* 84:5a, tr. by B. Watson, I, p. 505.

[4] The sentence is corrupt. 世人鏡鑒 reads 鑒乎前, 鑒乎人, 鑒乎鏡 in *I-lin* (*SPTK* ed.) 5:2b; also *CSCY* 46:6b.

[5] *Shun* 順 (observe, follow) reads *hsün* 訓 (lesson, teaching) in *CSCY* 46:7a. My translation combines these two versions.

[6] Founders of the Hsia and Shang dynasties.

learn from their people; wearing one's cap in an improper fashion and having a dirty face is due to the failure to look in a clear mirror. Therefore, a superior man should always make an effort to reflect upon things; but those who look only in the mirrors by their side will fail to see true reflections.

4.3 (1b5) Someone inquired about whether it is the monarch who embodies the essence of good government.

I said: "Everything has its complement. Were it not for [the complementary interaction of] Heaven and Earth, nothing would be born; were it not for [the cooperation between] the monarch and the ministers, good government could not be achieved. Heaven and Earth are the *sine qua non* for [the existence of] all beings, and monarchs and ministers are the rulers [of these beings]. This is the Way taught by the former kings. No one should ever transgress it, not even for a single moment."[7]

In the past, there were monarchs who became Sages through the teaching and admonition as well as the assistance they received from their ministers; as a result, the four neighboring peoples submitted respectfully [to them]. Therefore the office of Monitor-in-Attendance[8] should not be left vacant, the rituals and regulative ordinances should not be hidden from view, efforts to follow the teachings of the former Sages should not slacken, and the mind should not be exposed to improper ideas. As a result, that which is evil and perfidious will not have a way of infiltrating [one's mind]. If there is an opening, corrupting forces are sure to penetrate inside.[9] Once an evil idea is generated, evil conduct will follow and [the road to] justice will be obstructed. Once [the road to] justice is ob-

---

[7] Legge, I, pp. 383–384.

[8] *Chien-chia*, lit., seal scabbard, to frame, to restrain; *Hsün-tzu* (*SPTK* ed.) 4:22a; tr. by H. H. Dubs as "standards" in *The Works of Hsüntzu*, p. 118; *HHS* 49:33a.

[9] *Yu-chien*, a crack through which things leak in or out. According to Huang Hsing-tseng, there are two corruptions before and after the term. Huang's opinion, however, was contested by Lu in *Corrigenda* 2a.

structed, public-spiritedness and righteousness will not enter [one's mind].

Not to confide in one's favorites is just;[10] to follow only what is just is to be enlightened. Duke Huan of Ch'i was a man of moderate talent. The reason that he was able to accomplish meritorious deeds was due to an extraordinary [effort that he made]. In his harem he had many concubines, some of whom he might have loved more dearly than others; at his court he had many courtiers, some of whom must have been closer to him than others. [But among these concubines and courtiers, Duke Huan confided] only in Kuan Chung and Lady Wei—the former, an outsider who had once shot an arrow against the duke;[11] the latter, an old woman who had lost her charm.[12] This was not favoritism, but proper trust. From this we can learn [a moral]: trust only those who are virtuous and follow only those who are wise. How marvellous is this precedent!

If we keep the *kao-huang*[13] pure and clean, malignant elements[14] will not be generated there; this is called peace of mind. If [the monarch] keeps his imperial residence pure and clean, evil and perfidious persons will not find a place there; this is called peace of state. *Kao-huang* is a region so close to the heart that it can neither be reached by acupuncture needles nor be treated with medicine. Once it is affected by malignant

---

[10] *Jen*, to trust, to commission; *pu-ai*, not to favor, reads *so-ai* 所愛, the favorite, in *Hsiao Hsün tzu* 19a.

[11] For a description of the rise of Duke Huan to power in 686–685 B.C. and his relationship with Kuan Chung, see Legge, V, pp. 81–84. A more detailed account may be found in *Kuan-tzu* 7:1–10, tr. by W. A. Rickette, *Kuan-tzu* (Hong Kong University Press, 1965), pp. 45–67.

[12] For Duke Huan's favorite concubines in the harem, see Liu Hsiang, [*Ku*] *lieh-nü chüan* (*SPTK* ed.) 2:3, tr, by Albert R. O'Hara, *The Position of Women in Early China* (The Catholic University of America Press, 1945), pp. 50–52; also Legge, V, pp. 172–173. *Se-ch'ien* 色姜 (charming concubine) reads *se-shuai* 色衰 (charm declining in old age) in *CSCY* 46:7a.

[13] *Kao-huang*, the region between the heart and the diaphragm. For the medical significance of this part of the body, see Legge, V, pp. 372–374.

[14] *Erh-shu*, two page boys, referring to the disease which took the form of two boys in the dream of the Marquis of Chin in 580 B.C., *ibid.*

elements, a chronic disease develops. Let those who care for the human body or for the state beware this terrible [affliction].

4.4 (3a3) Someone said: "To love one's people as one's own sons—is this the highest degree of kindheartedness (*jen*) [attainable]?"
I said: "Not quite."
Someone said: "To love one's people as dearly as oneself—is this the highest degree of kindheartedness (jen) [attainable]?"
I said: "Not quite. King T'ang offered himself as a human sacrifice at Sang-lin;[15] [Duke Wen of] Chu [risked his own life in] moving his capital to I;[16] Duke Ching endured bodily suffering to pray for rain[17]—these may be called the Way to love one's people."
Someone said: "Why should a monarch love his people more dearly than himself?"
I said: "The sovereign receives the Mandate of Heaven to care for the people; so long as the people survive, the altar of the state survives. Once the people perish, so does the altar of the state. To care for the people is to care for the altar of the state and to obey the Mandate of Heaven."[18]

4.5 (3b9) Someone inquired: "Mencius said that everyone can become a King Yao or King Shun;[19] is that possible?"
I said: "So long as one is not a moron, it is possible to become

---

[15] According to legend, under the Shang dynasty the country suffered a great drought for five (another version alleges seven) years, whereupon King T'ang offered himself as a sacrifice to pray for rain. *Huai-nan tsu* 9:4a.

[16] Duke Wen of Chu consulted the oracle about moving the capital. It indicated that such an action would benefit the people, but harm the ruler. Considering the people more valuable than himself, he moved the capital and died soon afterwards. See Legge, v, pp. 263–264.

[17] Duke Ching of Ch'i, upon Yen Tzu's advice, lived without shelter under the burning sun, praying for rain; *Yen-tzu ch'un-ch'iu* (*SPTK* ed.) 1:18. See also *Shuo-yüan* (*Han-Wei ts'ung-shu*, First Series) 18:6b–7a; and Edward H. Schafer, "Ritual Exposure in Ancient China," *HJAS* 14 (1951), 130–184.

[18] Cf. Legge, II, pp. 300–301, 355–357, 483–484; v, p. 264.

[19] Legge, II, p. 424. Yao and Shun were legendary Sage-kings of antiquity.

a second Yao or Shun. It is not possible to bear a physical resemblance or to have a surname exactly the same as Yao's or Shun's; but it is possible to follow the rules or to practice the Way of Yao. Those who practiced the Way in the past emerged as the ancient Yao and Shun; he who practices the Way now will become a Yao or Shun of our times."

Someone said: "Is it possible for everyone to become a King Chieh or a King Chou?"[20]

I said: "Those who do the deeds of Chieh and Chou are like Chieh and Chou. The deeds of Yao and Shun, or Chieh and Chou, are always possible in this world. Men can choose [their paths of action]. Yang Chu wept at the crossroads because he was confronted with a difficult choice.[21] But why should one be saddened by a crossroads? A man may pick the middle road or turn on his heel. Although the alternatives may be as narrow and hazardous as the choice between the *Hsüan-tu* mountain paths,[22] a man should still follow his chosen way with composure."[23]

4.6 (4b2) Auspicious or inauspicious omens are subtle and yet clear. [The Viscount of] Chao was filled with concern at his dining table after [he found out that his army] had captured two cities;[24] the concubines of Master Chu of T'ao wept with deep sorrow [when they found out] that their husband had

---

[20] Chieh and Chou, wicked last rulers of the Hsia and Shang dynasties respectively.

[21] *Huai-nan tzu* 17:14b.

[22] *Hsüan-tu*, steep mountains in southwestern China that can only be crossed by tying oneself to a rope; *HS* 96A:9a.

[23] *Su*, common, usual; Legge, I, p. 395. According to Huang's commentary, to climb the *hsüan-tu* mountains refers to the hazardous Way of wicked rulers like Chieh and Chou. However, since *su* carries a strong connotation of virtue—such as in *su-wang* (the commoner-as-king, i.e., Confucius, the uncrowned prince), *su-chih* 素志 (one's cherished ideal), and *su-ssu* (untinted silk)—Huang's interpretation is misleading.

[24] The viscount was Chao Hsiang-tzu. When he heard that his army had captured two towns from the Ti people, he was worried that his people were not virtuous enough to deserve the victory. See *Kuo-yü* 15:8b.

THE SHEN-CHIEN

become rich.²⁵ A lesson can be learned from the above: what appears to be beneficial might turn out to be injurious and vice versa.²⁶

The cycles of submission and domination²⁷ are obscure and yet obvious. [King Shao-k'ang's] state of distress [when he was at] Yu-jeng signaled the restoration of the Hsia dynasty;²⁸ the omen of a pheasant [alighting on the handle of] a tripod presaged the resurgence of the Yin dynasty;²⁹ the tragedy at the Mansion of Shao augured the revival of the glory of the Chou dynasty;³⁰ [King Kou-chien's] stay in K'uai-chi paved the way for the state of Yüeh to achieve hegemony;³¹ the revolt staged during Tzu-chih's reign portended the emergence of the Yen as a strong state.³² From the above, a lesson can be learned:

---

²⁵ After Fan Li had helped the King of Yüeh restore the state, he left the government and changed his name to Master Chu of T'ao. He became a trader, but each time he became rich he would squander his money, quit his home, and start over again somewhere else. See *Shih-chi* 41:10b–14, tr. by Edouard Chavannes, IV, pp. 439–447; also *Shih-chi* 129:4, tr. by N. L. Swann, *Food and Money*, pp. 425–426; and by B. Watson, II, p. 481.

²⁶ The character 之 may be a corruption of 夕, indicating the repetition of the preceding character 損; cf. *Hsiao Hsün-tzu* 20a.

²⁷ Lit., bending and stretching.

²⁸ According to traditional records, the Hsia dynasty suffered a setback in 2118 B.C., when the fifth king was murdered and his queen gave birth to Shao-k'ang in exile at Yu-jeng. In 2077 B.C., Shao-k'ang restored the glory and power of the dynasty. See Legge, III, "Prolegomena," pp. 120–121.

²⁹ The incident occurred in the time of King Wu-ting, the 22nd ruler of the Shang dynasty. Because of this omen the king made an effort to strengthen his moral virtue, and the power of his dynasty grew. See Legge, III, pp. 7, 264–267; also *Shih-chi* 41:2–7.

³⁰ The Mansion of Shao was the residence of Duke Shao of Chou. A Chou prince sought refuge there after his father, King Li, was exiled in 842 B.C. The prince later became King Hsüan (r. 827–782 B.C.), who effected a mid-dynasty restoration of the Chou. See *Kuo-yü* 1:6b.

³¹ In 494 B.C., King Kou-chien of the state of Yüeh was defeated by the state of Wu. He fled to Mount K'uai-chi and humiliated himself in submission to the victorious state, while secretly preparing for revenge. In 473 B.C. he defeated and destroyed Wu and became a leading power among the rival states. See Legge, V, pp. 792–794; *Kuo-yü* 19:1–2, 6b–7a, 12b–18a; 20:1–5a.

³² Tzu-chih was a minister of King Kuai of Yen. He usurped the king's power

a period of submission may herald a period of domination and vice versa.[33]

4.7 (5b1) A sovereign always finds himself in a dilemma: either he causes trouble by remaining aloof and letting the state fall into disorder, or he troubles himself by taking an active part in government, toiling his body, straining his thought, and controlling his emotions in order to follow the Way. The [kind of] trouble that begets more troubles is chosen by unenlightened rulers; the [kind of] trouble that cuts short further troubles is chosen by those who are enlightened.[34]

A minister, too, finds himself in a dilemma: either he commits a crime by neglecting what loyalty and honesty require of him in office, or he becomes too dutiful and upright and ends up offending both his sovereign and his subordinates. The crime that is a real crime is committed by a crooked official; the second offense, which is actually not a crime, is preferred[35] by an honest official.

A minister's duty is such that he must not say, "My ruler is capable and does not need me; anything I say will be of no use," and thereby fail to do what loyalty demands. He must not say, "My ruler is not capable and cannot appreciate me; anything I say will be of no use," and thereby fail to do what loyalty demands. A man must do his best to comprehend the Way and to fulfill his obligations without swerve. Otherwise, he should retire from office and take care of himself [instead]. This is what the Way requires of a minister. Therefore, between a sovereign and his ministers there may be disagreements, but no conflict;

---

and prosecuted the prince heir-apparent, causing internal strife and invasion from the state of Ch'i in 314 B.C. Two years later the prince heir-apparent was installed as the new king. He achieved many reforms and defeated Ch'i. See *Chan-kuo t'se* (*SPTK* ed.) 9:12–13, 9:16–21, tr. by J. I. Crump, Jr., in *Intrigues: Studies of the Chan-kuo-t'se* (University of Michigan Press, 1964), pp. 65–66.

[33] 之 perhaps should read 夂 ; see note 26. This last sentence is not in *Hsiao Hsün-tzu* 20b.

[34] Cf. *Shuo-yüan* 16:12b.

[35] *Chih* (posit) reads 致 (effect, attain) in *CSCY* 46:8b.

there may be unpleasant occurrences, but no deep resentment; there may even be instances of yielding, but no humiliation.

4.8 (6b3) There are three possible crimes a minister may commit: first, misguiding [a ruler]; second, flattering a misguided [ruler]; and third, coveting improper favors [from such a misguided ruler]. To lead[36] a ruler into evil-doing is what I mean by misguiding; to follow a ruler into evil-doing is what I mean by flattering; to refrain from giving good counsel upon seeing [a ruler] misguided is what I mean by coveting favors. Ministers who misguide [a sovereign] should be executed; those who flatter [him] should be punished; those who covet favors from [him] should be demoted.

There are three ways a minister may proffer good counsel: by cautioning, by rectifying, and by reprimanding. [Giving good counsel] in advance is what I mean by cautioning; [giving counsel] to stop evil-doing[37] is what I mean by rectifying; giving reproof after an act has already been committed is what I mean by reprimanding. Cautioning is the best policy, rectifying is next, and reprimanding is the worst.[38]

When those in lower positions do not seal their lips and those in the highest positions do not block their ears, good counsel will be heard. It is possible to break a visible seal, but dealing with those that are not visible is truly a difficult [task]; it is possible to dissolve a block that one can see, but how is one to deal with those that cannot be seen?

4.9 (6b4) Someone said: "Should a sovereign ever yield?"

I said: "A sovereign should both assert himself and yield according to the precept of righteousness. Emperor Kao-tsu, although capable of establishing military domination over the regimes of Ch'in and Ch'u, yielded [when confronted by] the

---

[36] *Yin* (to lead) reads 先 *hsien* (to precede) in *CSCY* 46:8b.

[37] *Chih-chih* (to stop it) reads 進諫 *chin-chien* (to remonstrate) in *CSCY* 46:8b. My translation is a compromise between the two versions (*CSCY* in brackets).

[38] Cf. *Shuo-yüan* 2:1–3a.

Four Elders of Mount Shang;[39] Emperor Kuang-wu, although capable of delivering [crushing blows to Wang] Mang's regime, yielded to the stiff-necked magistrate [of Lo-yang];[40] Emperor Ming, although capable of commanding obedience throughout the empire, yielded to his Master-of-Writing Chung-li [I].[41] [On the other hand], the second emperor of Ch'in gave rein to his desires and ridiculed [the Sage-rulers of] T'ang and Yü;[42] the dowager Empress Ting-t'ao, née Fu, indulged in self-aggrandizement and caused grievance to the Cheng.[43] These are instances of unyielding [behavior which brought harm rather than good]. Otherwise, the dowager Empress née Chao[44]

[39] Emperor Kao-tsu (r. 202–195 B.C.) wanted to depose his heir-designate despite the strong objections of his ministers. He was dissuaded, however, when the Four Elders of Mount Shang appeared at the side of the heir-designate (the Four Elders were famous recluses who had refused to serve the emperor). The confrontation was recorded in *Shih-chi* 55:12b–13a, tr. by B. Watson, I, pp. 148–149.

[40] Tung Hsüan, Magistrate of Lo-yang, once offended an imperial princess in the course of carrying out his official duties. When the emperor ordered him to apologize by prostrating himself before her, he refused even when forced to the ground by two eunuchs in attendance. The emperor then forgave the "stiff-necked" magistrate for his unbendable sense of righteousness. See *HHS* 66:8b; 77:3b–4.

[41] Chung-li I several times gave strong remonstrance to the emperor, thereby changing the ruler's conduct. See *HHS* 44:15–17a.

[42] The second emperor of Ch'in considered himself a greater ruler than the Sage-kings Yao and Shun. He eventually caused the downfall of the Ch'in in 207 B.C. See *Shih-chi* 6:35b–36, tr. by Edouard Chavannes, II, pp. 207–209.

[43] Cheng was the surname of Lady Fu's stepfather. Lady Fu was a concubine of Emperor Yüan (r. 48–33 B.C.) of the Former Han dynasty. His successor, Emperor Ch'eng (r. 32–7 B.C.), had no heir. Lady Fu bribed the Empress née Chao to have Lady Fu's grandson, the Prince of Ting-t'ao, appointed as heir-designate. When he became Emperor Ai (r. 6–1 B.C.), Lady Fu became the Grand Empress. She ennobled many of her paternal relatives and even had a grandson of her stepfather made a marquis. When Wang Mang took over control of the imperial court in A.D. 1, she was posthumously disgraced and her relatives were exiled. See *HS* 97B:19–24a.

[44] See note 43. Because of her aid in enthroning Emperor Ai, the dowager Empress née Chao was well-treated by Lady Fu. But after Wang Mang came to power, he demoted her to the rank of a commoner and forced her to commit suicide. See *HS* 97B:11–18.

would not have been ruined and the Ch'in regime would have been unassailable. Therefore, a sovereign must assert himself or yield according to [the precept of] righteousness. By doing so, his cheerfulness will be like the sunshine in spring; his anger, like the frost in autumn; his authority, like the awe-inspiring thunderbolts; and his kindness, like the moistening rains and dewdrops—having an all-pervasive influence that none can resist."

4.10 (8b1) Someone said: "This is too difficult to practice."

I said: "Of course it is difficult for one to be like Emperor Kao-tsu, who, upon the advice of a conscript, immediately abandoned his favorite abode and moved his capital without a single day's delay;[45] or to be like Emperor Hsiao-wen, who refused to take possession of a horse that was able to traverse one thousand *li* [in a day], and [whose favorite] Lady Shen never wore gowns that trailed on the ground;[46] to be like Emperor Kuang-wu, whose hand never held pearls or jade.[47] But without mastering one's emotions and gaining control over one's desires, it is impossible to accomplish worthy deeds. It is also difficult to be a minister like Chin Mi-ti, who killed his own sons because of their unruly behavior;[48] or to be like Ping Chi, who never claimed merit for [his good works];[49] or to be like Su Wu, who persevered in his mission [abroad]."[50]

[45] Emperor Kao-tsu, founder of the Former Han dynasty, originally wanted to establish his capital at Lo-yang, former capital of the Eastern Chou. But Lou Ching, a conscript, warned the emperor that his moral influence could not equal that of the Chou rulers and advised him to establish his capital at Ch'ang-an, near the capital of the Ch'in dynasty instead. His advice was immediately accepted. See *Shih-chi* 99:1–3a, tr. by B. Watson, I, pp. 285–288.

[46] *HS* 4:15a, tr. by H. H. Dubs, I, pp. 272–273; 64B:15b–16a.

[47] *HHS* 76:1b.

[48] Chin Mi-ti, a sinicized barbarian who served as Emperor Wu's bodyguard. He executed his sons who had played improperly with a lady of the palace. See *HS* 68:21b.

[49] Ping Chi, as the officer in charge of the metropolitan prison, had saved the life of an infant who later grew up to be Emperor Hsüan (r. 73–49 B.C.); but he did not claim merit for the deed. See *HS* 74:8–10a.

[50] Su Wu, Han ambassador to Hsiung-nu, was detained for nineteen years (100–81 B.C.) without yielding to the enemy. See the biography in *HS* 54:16–21.

4.11 (9a9) Someone inquired about self-mastery.

I said: "King Kao-tsung of Yin was able to mend his ways and endure the most distressing medicine;[51] Duke Wu of Wei, [in spite of his old age of ninety-five], issued a proclamation at his court [encouraging his subordinates to criticize him];[52] King Kou-chien [of Yüeh] always kept a gallbladder hanging near his seat [as a reminder of his bitterness against the King of Wu].[53] How great these efforts at self-mastery are!"

4.12 (9b7) To be the beloved wife or favorite concubine [of a ruler], how fortunate! But what great calamities it may bring about! Is good fortune the same as calamity?

My answer is: Cultivating favors in a proper way leads to blessings; otherwise, [it leads to] calamity. Had Lady Ch'i not enjoyed the favor [of Emperor Kao-tsu], she would not have become a human swine;[54] had Lady Chao not enjoyed the favor [of Emperor Ch'eng], she would not have lost her life;[55] had Lady Li not enjoyed the favor [of Emperor Ching (r. 156–141 B.C.)], she would not have fallen into disgrace;[56] had Lady Kou-i not enjoyed the favor [of Emperor Wu (r. 140–86 B.C.)], she would not have died of grief in the prime of her life.[57] Are these not calamities? [On the other hand,

[51] Kao-tsung was the temple name of King Wu-ting (r. ca. 1334–1266 B.C.), the 22nd ruler of the Shang dynasty. According to the *Book of Documents* (*Shu-ching*) he was deeply concerned about his moral character as a ruler. He selected Fu Yüeh, a humble laborer but a worthy man, as his minister and asked Fu Yüeh to give him stern criticism—"like strong medicine which, if it does not distress the patient, will not cure his sickness." See Legge, III, pp. 248–253.

[52] Legge, IV, "Prolegomena," p. 76, pp. 510–518; *Kuo-yü* 17:11b–12a.

[53] King Kou-chien was said to have always had the gallbladder of a pig hanging in front of his seat. He would taste its bitter bile as a reminder of his humiliation by the King of Wu. See *Shih-chi* 41:3b, tr. by Edouard Chavannes, IV, p. 424. See also note 31 above.

[54] After Kao-tsu died, the dowager Empress née Lü ordered Lady Ch'i to have both arms and legs amputated, thus condemning her to live as a human swine (*jen-chih*). See *Shih-chi* 9:2b–3a, tr. by Edouard Chavannes, II, pp. 408–410; also B. Watson, I, pp. 322–323.

[55] See note 44.

[56] Both Lady Li and her son, the prince heir-apparent, fell victim to intrigue at the palace. See *Shih-chi* 49:8–9a, tr. by B. Watson, I, pp. 387–388.

[57] Lady Kou-i's son was appointed as the prince heir-designate (he later

there are examples of cultivating favor in the proper way] like the wisdom shown by Lady Shen,[58] the sagacity displayed by Pan *Chieh-yü*,[59] or the gracious virtue of Empress Ming-te[60] —how marvellous [they were]!

4.13 (11b2) He who makes the cares and joys of the world his own is a superior man; he who does not is an inferior man. In an age of great peace, [a superior man will have] few [cares about] worldly affairs, and all the people will be happy.[61]

4.14 (11b5) To ask a busy man [to follow the rules of etiquette by] bowing courteously and prostrating himself a hundred times is not [the proper way to enforce observance of the] rites. To ask someone who is grief-stricken to sing to [the accompaniment] of the strings and to play the stringed instrument (*se*) is not [the way to manifest the true nature of] music. [The essence of proper observance of the] rites is reverence; [the essence of the true nature of] music is harmony. [In this sense], even the ordinary man and woman working in the paddy-fields should be able to observe the rites and [enjoy] music.

4.15 (11b9) Those who disagree with the sovereign so as to follow the Way are loyal ministers; those who deviate from the Way so as to agree with the sovereign are ministers of flattery. By being loyal, one benefits the sovereign; by flattering, one benefits himself. Loyal ministers find comfort in their minds; ministers of flattery find comfort in their bodies. Therefore,

---

became Emperor Chao, r. 86–74 B.C.). Emperor Wu feared that Lady Kou-i was too young to be the mother of a future emperor and ordered her to be executed under a pretext. See *HS* 97A:17.

[58] She used to sit improperly before the emperor and empress, but mended her ways at the injunction of a scholar-official. See *HS* 49:3.

[59] A woman scholar. See *HS* 97B:8–11a; also N. L. Swann, *Pan Chao*, p. 26.

[60] Another woman scholar, who wrote the *Hsien-tsung ch'i-ch'ü chu* (*Daily Records of Emperor Ming*). See *HHS* 10A:11b–19a.

[61] Cf. Legge, II, p. 158.

a sovereign must notice the way his ministers agree or disagree [with him], study the motives behind their actions, and examine what comforts they would find therein. Prince Kuang-ch'uan did not heed this [advice] and consequently executed his loyal ministers;[62] King Kung of Ch'u took heed too late, and so uttered words [of remorse] on his deathbed;[63] King Hsüan of Ch'i apparently took heed, and therefore rewarded those who remonstrated with him.[64]

4.16 (12b5) An inquiry was made about the precautions a sovereign and his ministers should take.

I said: "Precautions should be taken against almost everything." When I was asked to elaborate on the most important [matters about which to be concerned], I said: "A sovereign should exercise caution against too much self-indulgence; a minister, against too much self-interest."

4.17 (12b6) (*CSCY* 46:8b9–9a9)

There was a question: "[Should] the son of Heaven defend his borders against the four barbarians?"[65]

I said: "This constitutes only [a means of] external defense. The [means of] internal defense for the son of Heaven lies in his own body."

---

[62] *HS* 53:15b–16.

[63] King Kung of Ch'u neglected the advice of his loyal ministers and lost a battle with the Chin in 574 B.C. He publicly acknowledged his regret on his deathbed in 559 B.C. See Legge, v, pp. 389–398; 456–458.

[64] *Chan-kuo t'se* 4:12–17a; J. I. Crump, Jr., *Intrigues*, pp. 2–4.

[65] There is a major corruption in this passage. The amendment is from *CSCY* 46:8b9–9a9, which reads:

或問天子守在四夷，有諸？曰：此外守也。天子之內守在身。曰：何謂也？曰：至尊者，其攻之者衆焉。故便僻御侍攻人主而奪其財；近幸妻妾攻人主而奪其寵；逸遊伎藝攻人主而奪其志；左右小臣攻人主而奪其行；不令之臣攻人主而奪其事；是謂內寇。自古失道之君，其見攻者衆矣。小則危身，大者亡國。鯀共工之徒攻堯；儀狄攻禹，弗能克；故唐夏平。南之威攻文公；申侯伯攻恭王，不能克；故晉楚興。萬衆之寇凌疆場，非患也。一言之寇襲於膝下，患之甚矣。

See also *Hsiao-wan-chüan lou ts'ung-shu* edition, 4:5a7–5b6.

He said: "What is the meaning of this?"

I said: "When a man is in the highest position of power, those who attempt to assault him will be many. His personal attendants and bodyguards might attack him and steal his property; his closest companions [or his] wife and concubines might attack him and try to win his favor; lazy, shiftless artists and entertainers might attack him and destroy his determination; petty officials around him might attack him and corrupt his moral conduct; treacherous officials might attack him and ruin his affairs—these are the internal bandits. From times of old there have been monarchs who have lost the Way and were besieged by a host of such bandits. In minor cases [of this sort], the life of a monarch was at stake; in major cases, a dynasty was destroyed as a consequence. K'un, Kung-kung, and their followers besieged King Yao,[66] and I Ti besieged King Yü;[67] but their attempts failed and the kingdoms of T'ang and Hsia enjoyed peace. Nan-chih-wei besieged Duke Wen,[68] and Shen, the grand marquis, besieged King Kung;[69] they also failed, and the states of Chin and Ch'u rose to power.

"An invasion of the frontier by a legion of ten thousand

---

[66] As the legend goes, because these officials were fine talkers but could not administer the affairs of state they were executed by King Yao. See Legge, III, pp. 23–25, 39–40.

[67] I Ti made some very delicious wine for the pleasure of King Yü, ruler of the Hsia dynasty, but the king turned it down out of fear that indulgence would lead to the ruin of the state. See *Chan-kuo t'se* 7:7b.

[68] Nan Chi-wei, a beautiful girl who charmed Duke Wen of Chin and made him stay away from court for three days. He later deserted her because he feared her strong attraction; *ibid.*

[69] Shen was the surname of a powerful clan of the state of Ch'u during the Ch'un-ch'iu period. Duke Wu-ch'en of Shen voiced objection to the infatuation of King Kung (r. 590–559 B.C.) with Lady Hsia, a captive from the state of Ch'en, and also admonished several Ch'u nobles for their intimacy with her (588 B.C). See Legge, V, pp. 341–347. Later Tzu-chung and some other nobles exterminated Wu-ch'en's family. Wu-ch'en then decided to become a supporter of the state of Chin and led the allied armies of Chin and Wu in an attack on Ch'u (583 B.C.). See Legge, V, pp. 362–364. It is also recorded that in 570 B.C. Tzu-chung executed another Duke of Shen, Marshal of the Right (*Yu ssu-ma*), who was said to be very corrupt and oppressive. See Legge, V, pp. 414–415, 417.

bandits is not a real menace; far more dangerous is a sentence [uttered by one evil person kneeling] at the sovereign's feet. A gift presented by [an ambassador from] one of the eight foreign states using a language that needs extensive translation should not be regarded as something priceless; whereas good counsel offered by a close supporter as he prostrates before the monarch is a truly priceless treasure. Therefore, an enlightened ruler should pay great attention to internal defense—ridding himself of internal bandits and prizing the internal treasure of good counsel."

4.18 (12b8) The clouds move with the dragon; the wind follows the tiger.[70] The phoenix danced elegantly to the *shao* music;[71] the unicorn came to Confucius.[72] These are [examples of] reciprocal phenomena—the one appears, the other responds. Good behavior brings about blessings; blessings bring about good fortune. On the other hand, evil actions bring about evil omens; evil omens bring about bad luck. Therefore a superior man responds to [the appropriate phenomena].

4.19 (13a7) A superior man follows a harmonious diet to regulate his energies, listens to harmonious sounds to regulate his mental attitude, accepts harmonious counsel to regulate his administration, and takes harmonious action to regulate his morality. Sourness, saltiness, sweetness, and bitterness differ in taste, but together they contribute to good flavor—this is what I mean by a harmonious diet. *Kung, shang, chüeh,* and *chih*[73] differ from each other, but together they contribute to good music—this is what I mean by harmonious sound. [Words of] approval or disapproval and criticism or suggestion differ from each other, but [following] the golden mean and [the precept of] justice should always be one's guiding principles—this is what I mean by harmonious counsel. Approach

---

[70] *I-ching* 1:15a, Legge, *SB*, XVI, p. 411.
[71] *Shao*, music of the legendary King Shun; Legge, III, pp. 88–89 and note.
[72] Legge, V, pp. 833–835.
[73] Ancient Chinese musical notes.

and retreat, activity and quietude, differ from each other, but balance between them contributes to the refinement of conduct —this is what I mean by harmonious action.

There is a saying: "If a man takes the words of others only on condition that they are not contradictory to his own, then the state will be ruined."[74] Confucius said: "The superior man [inspires] harmony but not adulation."[75] Yen-tzu also said: "If one prepares food by flavoring water with water, who can bear to eat it? If one plays a monotone on the stringed instruments of *ch'in* and *se*, who can bear to listen to it?"[76]

The ode reads:
> There are also the well-seasoned soups,
> Carefully prepared in advance, the ingredients
>   rightly proportioned.
> By these offerings we invite His presence,
>   without a word,
> Nor is there now any contention.[77]

[This] alludes to the same [idea].[78]

---

[74] Legge, I, p. 269.

[75] Legge, I, p. 273.

[76] The *ch'in* and *se* are ancient Chinese musical instruments. The *ch'in* has five or seven silken strings, and the *se* has metal strings of varied number. The quotation is from Legge, V, pp. 679–684, in which Yen-tzu presents a lengthy discourse on harmony.

[77] Legge, IV, pp. 634–635.

[78] Here Hsün Yüeh advocates a harmony of diverse elements rather than mere conformity.

Shen-chien 5
Miscellaneous Dialogues
(*Tsa-yen*), II

5.1 (1a4) He who wears [fine] clothes does not wallow in the dirt[1] because he cares about [his attire]. But how superficial is he who cares for his clothes but not his manners, how much worse is he who cares for his manner but not his speech and actions, and how shallow is he who cares for his speech and actions but not the enlightenment of his spirit![2] Therefore, a superior man fundamentally [considers] spirit (*shen*) to be that which is most precious. [To attain] harmony of spirit and balance in one's deportment in order to gain access to the Way is to protect one's true [nature].[3]

5.2 (1a9) Man builds his virtue out of three things: first, integrity; second, understanding; and third, determination.[4]

[1] *Ch'en-t'u* may also mean a dusty road. My translation follows Huang's commentary.

[2] Cf. *Han-shih wai-chuan*, tr. by Hightower, p. 32. The term *ming* (enlightened, discriminative comprehension) usually refers to the intellect or mind of an individual. But it is also frequently used in combination with *shen* (spirit, also of an individual) to form the term *shen-ming* which refers to the Divine Intelligence. The spirit of an individual thus implies an endowment by Heaven, Nature, and the Universal Spirit.

[3] As discussed in the preceding note, the term *shen* (spirit) implies both the divine spirit and the human spirit (the moral/intellectual essence of the mind of an individual). The notion that man's spirit is endowed by Heaven, Nature, and the Universal Spirit was shared by both the Confucians and the Taoists. The Confucians emphasized cultivating and developing man's spirit to the utmost; the Taoists emphasized preserving it in, or restoring it to, its original "true" state. Hsün Yüeh's precept of "protecting one's true nature" is thus akin to the Taoist conception.

[4] *Chen*, decent and incorruptible; *ta*, lit., arriving, attaining; *chih*, ambition, ideal.

Integrity is the essence [of virtue]; understanding is [the means] by which to achieve it; and determination is [the power] to accomplish it. [He who possesses these] is a superior man. If a man is not able to possess all three [qualities] but only one of them, [let him seek] the most important: integrity.

5.3 (1b3) Man achieves self-discipline in four [areas]: sincerity of mind, uprightness of intention, truthfulness of action, and firmness in [adhering to] his assigned position. When his mind is sincere, even the Divine Spirit[5] will respond to him, not to mention the myriad people. When he is upright in intention, even Heaven and Earth will be in harmony with him, not to mention the myriad things. When he is truthful in his actions, he will accomplish meritorious deeds. When he is firm in his position, he will not go astray.

5.4 (1b6) Someone said: "[A man's] ability (*ts'ai*) is the substance [of his being].[6] [A man] may alter [his] conduct, but [he] cannot alter [his innate] ability."[7]

I said: "In ancient times, what was called ability referred to essence; now what is called ability refers to the end-product.[8]

---

[5] The Chinese term is *shen-ming*; see note 2.

[6] The Chinese term for ability is *ts'ai* 才, the ability and capacity to act (to do things). It is the cause or at least the potentiality for action, and therefore exists prior to action. It is in this sense that *ts'ai* is considered to be an essence, unaffected by human action. The philosophical issue has been raised regarding whether *ts'ai* is inborn or acquired, and whether it is concerned purely with the capacity of one's intellect. During the Wei and Chin periods, it was widely believed among the cultural élite that *ts'ai* was inborn, predetermined, and possibly inherited, and that it could not be acquired by individual effort. This led to a dispute about the relationship between *ts'ai* and *hsing* (man's moral nature), which the Confucians also considered to be inborn. The subject was hotly debated among the *ch'ing-t'an* (Pure Conversation) groups of the time. See Hsün Yüeh, pp. 144, 167, 170, and notes. See also note 8 below.

[7] The sentence may also mean: "[A particular] thing can be done, but one's [limited] ability cannot do it." For Mencius' discussion of the potential of being virtuous and the capacity to actualize that potential, see Legge, II, pp. 402–403.

[8] The relationship between *ts'ai* and *hsing* is complicated by their archaic

Thus when [moral] conduct is exalted, one's ability is not put to improper use. But ability as such can be abused. There is an old story: A man who wants to go to the state of Ch'u[9] points his chariot toward the north, saying: 'My horses are good, my provisions bountiful, and my driver well-trained.' [He does not realize] that the more splendid these three, the farther he might travel away from Ch'u.[10] A man must lead [his horses] along [the proper] road and move in the [right] direction. When a superior man practices good conduct, he will accomplish it."[11]

---

and later (late classical and Han) meanings. The later meaning of *ts'ai* as ability implies that it can be acquired by training—hence the connotation of an end-product. However, *ts'ai* also has the archaic meaning as the material, the basic stuff of which things are made (the derivative *ts'ai* 材, the wood which is the basic stuff of all wooden objects); hence the connotation of essence. *Hsing* has the later Confucian meaning of "the moral good, or morally conditioned nature of man"; since some have argued that moral goodness is the result of training, it may imply that the so-called human nature is the end-product of proper training or cultural conditioning. This issue divided the Confucians between the followers of Hsün Tzu and the followers of Mencius. But *Hsing* also has an archaic meaning, the essence of life (生 in archaic script); see Fu Ssu-nien, *Hsing-ming ku-hsün pien-cheng*, also Wolfgang Bauer, pp. 36–38. Hence the important debate over whether *ts'ai* is identical to *hsing* as the essence of human life. See note 6.

In this passage Hsün Yüeh distinguishes between the archaic and later meanings of *ts'ai*, referring to the former as essence and the latter as end-product. It is in the latter sense that *ts'ai* is used in the discourse in this paragraph and the next. Hsün Yüeh also avoids using the term *hsing* 性 with the meaning of inborn essence in man, employing instead *hsing* 行 (moral conduct), which is clearly external and observable. The meanings of *ts'ai* (ability) and *hsing* (conduct) can thus be clearly differentiated.

[9] A feudal state in south China during the Spring-and-Autumn and the Warring States periods.

[10] *Chan-kuo t'se* 7:61.

[11] Note the character *chih* 至 (arriving at, attaining, utmost), which appears in *SC* 5.10 and recurs in 5.11, 5.15, 5.24, and 5.25, with the meaning of superlative ("the three superlatives"). Thinkers in the late Han and Six Dynasties periods became increasingly concerned about the discrepancy between knowing an ideal (such as "truth," "moral goal," the Way), and realizing (actualizing, reaching, and becoming one with, *chih*) it. They regarded mere knowing without realizing as superficial, and yet considered perfect realization to be

5.5 (2a4) Someone inquired: "The Sages were exalted; was it on account of their ability (*ts'ai*)?"[12]

I said: "[When ability] is used in combination with good conduct, it (ability) should be prized; but when it is not joined to good conduct, one should prize good conduct. If a man has the ability of Kung Shun and King Yü[13] and does no evil, he will be more...."[14] [If a man is] as kindhearted as Shun and Yü but does not have the same ability, he will still be a good man."

5.6 (9a9) Someone inquired about whether it is more difficult to give remonstrance or to accept it.

I said: "Lately it is difficult to give remonstrance because it is difficult [for one] to accept it. If [a man] has no difficulty in accepting it, then it will be easy [for others] to give it."

5.7 (6b3) Someone inquired about whether it is more difficult to know others or to know oneself.[15]

I said: "To know oneself, the search is inward and [begins] close at hand; to know others, the search is outward and [begins at] a distance. To know others is more difficult. If [you want to understand] all that can be known about this issue,[16] [it is that] one may become enlightened as to what lies within [oneself] through knowledge, while what lies outside is in the dark; and one may preserve what lies within

---

quite impossible, as shown in Yüeh Kuang's famous thesis on "a reaching (*chih*) that does not reach"—"he touched (reaching) the table with a fly wisk and then lifted it away (not reaching)." See Fung Yu-lan, *A Short History of Chinese Philosophy*, pp. 217–218.

[12] The sentence may also mean: "Was it ability that was most prized by the Sage?" For ability, *ts'ai*, see note 6.

[13] Legendary Sage-kings of antiquity.

[14] The text is corrupt. Hsün Yüeh seems to mean that such a man is much better than a man who is virtuous but of moderate ability.

[15] Cf. *Lao-tzu* 33, tr. by A. Waley, *The Way*, p. 184.

[16] *Shu*, numbers, fate, potentialities, the variegated elements in a given situation. Cf. *SC* 1, note 72.

[oneself] by concealment, while what lies outside is exposed.[17] Therefore, knowing others [is not the same as] knowing oneself. One may know oneself but not know others. How urgently [important is this matter]!"[18]

5.8 (9b9) It would be unusual if someone who was self-assertive did not do something abnormal. The superior man abhors three types of abnormal [tendencies]: the inclination to make trouble, the inclination to cause [something] unusual [to happen], and the inclination to deviate from the norm. If a man has an inclination to make trouble, he will bring many issues to the fore that will disturb the masses. If a man has an inclination to cause something unusual to happen, he will abandon the Way and confound the social customs. If a man has an inclination to deviate from the norm, he will slight the law and violate the rules. Therefore, when one [admires] fame one does not [also] respect notoriety; when one [admires] good conduct one does not [also] respect foolhardiness.[19]

5.9 (3a4) [Doing what is] expedient (*ch'üan*) may be a splendid [precept], but it is not as subtle as the permanent Canons (*ching*).[20] Sophisticated arguing may sound clever, but

[17] The meaning of this sentence is not clear. Huang's commentary reads: "One may be enlightened in knowing himself but blinded in knowing others; or one may be blinded in knowing oneself but enlightened in knowing others." This is not very convincing, for in another passage Hsün Yüeh asserts that one knows better in the dark (*SC* 1.43). See also A. Waley, *The Way*, p. 171. The character *ch'üan* reads 或 in the *SC* editions no. 6B, 6C, 6D, 8, 9, 10, and 11 (for these editions, see Introduction). This simplifies the problem, although not on authoritative grounds. *Ch'üan* (to keep, to save, self-preservation) is a very important concept in Hsün Yüeh's thought.

[18] The entire passage is quite corrupt. Huang's commentary on the last sentence reads: "To know others is more difficult, but to know oneself is more urgent." The explanation, although well-reasoned, does not seem to accord with the text. See Introduction, p. 51.

[19] *Han-shih wai-chuan* 3:18b, tr. by Hightower, p. 116. *Kou-nan*, [actions which are] unusually difficult and trying.

[20] *Ching*, constant, canonical, in contrast to *ch'üan*, circumstantial, expedient. See also *SC* 1, note 1.

THE SHEN-CHIEN

it may not be as well-reasoned as stammering speech.[21] That which is cultured and elegant may stand out, but it may not make its point as well as that which is plain.[22] [A man with] wide-ranging knowledge seems extraordinary, but [he] may not be as correct as [a man who is] concise.[23]

5.10 (3a7) There is no one who does not strive for that which is great; but only he who knows the essence[24] of the Way arrives at that which is greatest.[25] There is no one who does not seek after that which is mysterious, but only he who knows the nuances (*chi*) of the spirit (*shen*)[26] arrives at that which is most[27] mysterious. There is no one who does not desire that which is right; but only he who knows the ... of the ...[28] arrives at the peak of rectitude.[29] Therefore, the superior man must aim for the three superlatives.[30] [If he] fails [to attain] them, he must devote himself to [that which is proper and possible within his limited sphere of ability] and not transgress this.

5.11 (3b2) Someone inquired about [the writings to which one] should adhere.[31]

---

[21] *Cho* 絀, bending and yielding, may also read *ch'ü* 詘, stammering and hesitating in speech. Cf. *Lao-tzu* 5, 45, tr. by A. Waley, *The Way*, pp. 147, 198 (especially note 3).

[22] Cf. *ibid.*, 37, p. 188 (especially note 1).

[23] Cf. *ibid.*, 23, p. 171. 約 reads 傳 in edition no. 4. For the meaning of *yüeh* (concise) see *SC* 2, note 80.

[24] *T'i*, body, foundation. See Walter Liebenthal, tr., *The Book of Chao* (Peking, 1948), pp. 17–18.

[25] *Chih*, see note 11.

[26] *Shen*, see note 3. *Chi*, an omen, revealing the subtle mechanism (*chi*, pivot) by which the Way operates—a mechanism that can be foreseen by the initiated. See *I-ching* 7:25b, 8:13a, Legge, *SB*, xvi, pp. 370–392.

[27] *Chih*, see note 11.

[28] Two characters missing.

[29] *Chih*, see note 11.

[30] *Ts'un*, keep [in mind], cherish [the ideal]. The three superlatives (*chih*); see note 11.

[31] The question is a logical extension of the preceding dialogue.

THE SHEN-CHIEN

I said: "Only to the Sacred Canons. As for the Hundred Schools, [one cannot] adhere [to them]. [For of all the writings] there is none that does not have something to say, but only the essential [ones] are superlative;[32] there is none that does not point to what is moral, but only the mystical ones (*hsüan*) are really profound (*ao*);[33] there is none that does not follow the Way, but only [the Way] of the Sage is great. The Way of the Sage is the Way of the mean,[34] which reaches out in the nine directions."[35]

5.12 (3b8) Someone said: "Speech is nothing but [a means of] communication.[36] Why then is the speech of the Sage so cultured?"[37]

[I said: "The speech of the Sage is cultered][38] because it is profound in five respects: [it is] mystical, subtle, all-inclusive, concise, and cultured. [Because it is] abstruse and deep I call it mystical; [because it demonstrates] intricate reasoning I call it subtle; [because it displays] breadth of scholarship[39] I call it all-inclusive; [because it uses] few words I call it concise; [because it is a form of] accomplished exposition[40] I call it cultured. The polished writings of the Sage accomplish these five things. Therefore I said they cannot avoid [being cultured]."

5.13 (4a4) The superior man rejoices in his Heaven[ly-endowed nature] and knows his destiny (*ming*); therefore he has no anxiety.[41] He studies matters carefully and makes clear

---

[32] *Chih*, see note 11.

[33] *Hsüan*, darkness, mystical; *ao*, deep and abstruse.

[34] For the way of the mean, see Legge, I, pp. 382–389; Fung Yu-lan, *History of Chinese Philosophy*, I, pp. 369–377.

[35] *Ta = chih*, to reach, to attain; see note 11.

[36] Legge, I, p. 305. *Ta*, see note 35.

[37] *Wen*, see *SC* 3, note 12.

[38] This sentence is either the concluding part of the question or the beginning of the answer; see *SC* 1, note 65.

[39] *Shu*, numbers; see *SC* 1, note 72. *Shu-po*, lit., encompassing vast numbers.

[40] *Chang*, elaborating richly and clearly as in many chapters or sections.

[41] *I-ching* 7:10a, Legge, *SB*, XVI, p. 354.

judgments; therefore he is not confused. He settles his mind and does what is fair;[42] therefore he is not afraid. Nevertheless there is something he worries about and fears: he worries that he cannot fulfill [the potential of] his Heaven[ly-endowed] nature, and fears that he may be confused....[43] But worry cannot [help one to] escape what is destined by Heaven; and one must not be confused [by this].

5.14 (4a8) Someone inquired about human nature (*hsing*) and destiny (*ming*).[44]

I said: "That with which one is [endowed at] birth is called human nature, that is, body and spirit. That which [causes] the development and termination of life is called destiny, that is, good and bad fortune. At the moment of birth,[45] my nature and my destiny were conceived. The superior man accords with his nature so as to complement his destiny. He goes along with good fortune[46] and stands firm in the face of bad fortune as they may come his way, not striving after the one nor begrudging the other. But he who craves improper favors will pride himself upon [the good luck that is] destined by Heaven; and he who abhors bad [luck] will perversely defy what is destined by Heaven. Consequently, his pride will make him incapable of acting in harmony with [the good fortune that may come his way and he will fail to achieve] ultimate success; and his perversity will make him incapable of standing firm to the end. [He who] craves [improper favors] will become remiss; [he who] abhors bad [luck] will [make his situation] worse. Striving for good fortune and begrudging [fate], one begets calamity. This is to be confused."[47]

---

[42] *Chih-kung* also means "devote to the public."

[43] This passage is very difficult and may be corrupt at the place indicated by the ellipses. Note the pessimism, as discussed in note 11.

[44] The question is a continuation of the preceding passage.

[45] 制 *chih*, control, may refer to the factors conditioning one at birth; it may also be a corruption of 際 *chi*, at the moment.

[46] *Hsiu*, fortunate, reads 禮 *li*, rites, in edition no. 4.

[47] Cf. Lao-tzu's teaching on good and bad fortune. See A. Waley, *The Way*, pp. 152, 212.

5.15 (5a1) Someone inquired about [what things are] destined by Heaven and [what things can be changed by] human effort.

I said: "There are three classes: those in the upper and lower ranks cannot be changed, but those in the middle can be influenced by human effort."[48] [The sorts of things] that are destined [by Heaven] are similar [for all people], but the things that can be accomplished by human effort may vary greatly [among individuals].[49] This variation [in the efficacy of human effort results from] differences in the [dispensation of] good and bad fortune. Therefore [the *I-ching*] says: "By employing reason (*li*) and developing one's nature to the utmost, one arrives at [an understanding of] fate."[50]

Mencius said that human nature is good; Hsün Ch'ing said that human nature is bad; Master Kung-sun said that human nature is neither good nor bad; Yang Hsiung said that man's nature is a mixture of good and bad; Liu Hsiang said that since human nature corresponds to human emotions, the one cannot be all good and the other cannot be all bad.[51]

Someone asked: "What are the reasons [for these views]?"

I said:[52] "If human nature were [all] good, there would not

---

[48] For the ranking of personalities into three classes, see Legge, I, pp. 191, 313–314, 318; A. Waley, *The Way*, p. 193; A. Forke, *Lun-heng*, II, pp. 386–387.

[49] Legge, I, p. 318.

[50] From *I-ching* 9:3a. Here neither Legge's rendering in *SB*, XVI, p. 423, nor R. Wilhelm's in *I-ching*, p. 281, comes to the point of this specific reference to the *Book of Changes*. According to the context of the passage, the three things mentioned in the *I-ching* (*li*, *hsing*, and *ming*) refer to the three things discussed in the preceding sentence, i.e., what is destined by Heaven, what can be accomplished by human effort, and what is the result of good or bad fortune. Thus *li* (reason, law) means not only the law of nature but also what is ordained in man by nature. Here Hsün Yüeh's concept of *li* anticipates the Neo-Confucian identification of *li* (reason) and *hsing* (human nature). For "arrives at," see note 11. See also Legge, I, pp. 415–416.

[51] Legge, II, pp. 234, 394–403. Kung-sun tzu reads Kun-tu tzu in *Mencius*, Legge, II, p. 399–401. *Hsün-tzu* 17:1–16, H. H. Dubs, *The Works of Hsüntze*, pp. 301–317. Yang Hsiung, *Fa-yen* (*Han-Wei ts'ung-shu*, First Series) 2:3b. Wang Ch'ung, *Lun-heng* (*Han-Wei ts'ung-shu*, First Series) 3:16–21; A. Forke, I, pp. 384–391.

[52] It is not clear whether this is Liu Hsiang's opinion or Hsün Yüeh's.

have been the Four Criminals.[53] If human nature were [all] bad, there would not have been the Three Virtues.[54] If human nature were neither good nor bad, King Wen's teaching [to all his sons] being the same, there would not have been [such a difference between] the Duke of Chou, and Kuan and Ts'ai.[55] If human nature were all good and human emotions all bad, [you would have to say that] King Chieh and King Chou[56] had no nature at all, and King Yao and King Shun had no feelings.[57] If human nature were a mixture of good and bad, the highest Sage would have some evil in his mind and the lowest moron would have some good. These are not very well reasoned [notions]. But what Liu Hsiang says is correct."

---

[53] The four wicked ministers or barbarian chieftans in the time of the legendary King Shun. See Legge, III, pp. 39–40.

[54] The three men of virtue praised by Confucius. See Legge, I, p. 331.

[55] King Wen, founder of the Chou dynasty; Duke of Chou, the virtuous son of King Wen; Kuan and Ts'ai, two wicked sons of King Wen who later rebelled and were prosecuted by the Duke of Chou. See Legge, III, pp. 487–488 and note.

[56] The wicked last rulers of the Hsia and Shang dynasties.

[57] In this and the next three discourses, Hsün Yüeh's concepts of human nature (*hsing* and *ch'ing*) and moral judgment differ considerably from those of many later Confucians, and come close to the ideas of the Taoists and the Legalists. These Confucians postulated a polarity between *li* (reason) and *ch'i* (matter), between *hsing* (man's moral nature) and *ch'ing* (man's amoral feelings) or *yü* (man's amoral or immoral desires), between the rational and the irrational, and between the good and the bad. Hsün Yüeh refused to accept such a view. He suggested that *hsing* (human nature) and *ch'ing* (human feelings) are both aspects of man's essence, an essence endowed by Heaven and originating in Nature; and as such they are internal, undifferentiated, and morally indeterminate. This comes close to the Taoist view of nature and the original true state of man. Hsün Yüeh differed from the Taoists, however, in his insistence that the internal essence of man needs to be developed, externalized; and that a moral judgment must be made about the end-product—the deeds of man and the choices he makes in the outside world. This emphasis on judging the result rather than the motives of man's conduct, and attention to the external forces affecting action (outside attractions, etc.) and the importance of the governmental programs of education and law, resemble Legalism.

5.16 (5b7) Someone said: "Benevolence (*jen*) and righteousness (*i*) are [expressions of] human nature; likes and dislikes are [expressions of] human feelings. Benevolence and righteousness are always good, whereas likes and dislikes can sometimes be bad. Consequently, there are human emotions that are bad."

I said: "This is not so. Likes and dislikes represent the [choices of] acceptance and rejection that are made by one's nature. Because these [choices] are externally visible they are called human feelings; nevertheless, they originate from human nature. Benevolence and righteousness refer to good deeds that have already been accomplished;[58] therefore, how can they be other than good? Likes and dislikes arise before [the occurrence of specific acts] that one identifies as good and evil; therefore, why should one be surprised if some evils result [from such feelings]?

"Generally speaking, spirit[59] most closely resembles 'air.'[60] If there is 'air' then there is form.[61] If there is spirit, then there are feelings of like and dislike, happiness and anger. Therefore, feelings are to man's spirit as form is to 'air.' 'Air' can produce black or white;[62] spirit can produce good or evil. Therefore

---

[58] *Ch'eng*, lit., sincere, when used as a verb means sincerity not merely as an abstract principle or tendency but also in the form of proven or accomplished virtue (= 成 accomplished). Legge explained *ch'eng* as "real" in his note, II, p. 358. See also pp. 366–367, 413–419 (especially p. 418, "Sincerity is self-accomplishment and it accomplishes others").

[59] *Shen*, see notes 2 and 3.

[60] *Ch'i* in a broad sense means the "ethereal essence" that constitutes the substance of all things (*wu*). In a narrower sense, especially in vulgar Taoism, it refers to the air which one inhales and exhales, and was thought to be directly related to the vital fluid of the organs determining one's physical and mental condition. Hence the connotations of *ch'i* as air, vital fluids of life, and characteristics of one's personality or manners (as in *ch'i-chih*).

[61] *Hsing*, shape, form, appearance. See *SC* 3, note 17.

[62] The "ethereal essence" was thought to consist of the light-and-pure and the heavy-and-turbid, which determine the quality of all things; hence, the white (light and pure, *yang*) and the black (heavy and turbid, *yin*).

form is accompanied by [the appearance of] black or white, while feelings are accompanied by [the appearance of] good or evil. Hence, black 'air' is not the fault of one's form and an evil spirit is not the fault of one's feelings."

5.17 (6a8) Someone said: "Man's [likes and dislikes] are such that once he sees there is profit [to be had], he immediately likes it. He can be restrained by benevolence and righteousness [only] because his [moral] nature suppresses his feelings. When his [moral] nature is weaker than his feelings[63] and cannot suppress them, then his feelings will have their own way in doing evil."

I said: "It is not so. This (committing evil) occurs because [the attraction of] the good is weaker than the [temptation of] evil,[64] not because feelings [are themselves evil]. [For instance], if there is a man who likes both wine and meat, when the [attraction of] meat wins out, he will eat; when the [attraction of] wine wins out, he will drink. These two [attractions] contend with each other; that which wins will have its way. This is not because his feelings make him want wine and his nature makes him want meat. Now if there is a man who likes [to pursue] both profit and righteousness, when the [attraction of] righteousness wins out, he will seek righteousness; when the [temptation of] profit wins out, he will seek profit. These two (profit-seeking and righteousness) also contend with each other; that which wins out will have its way. This is not because his feelings make him want to seek profit and his nature makes him want to seek righteousness. When he is able to have both, he will take both; but when he cannot have both of them, he will take the most tempting one. If he likes only one of them, then even though he is able to take both, [he will not].[65] If he likes both the same and they are equally tempting to him, then

---

[63] Lit., "*hsing* are few; *ch'ing* are many."

[64] Lit., "This is [due to] the fewness and many-ness of good and evil."

[65] The sentence is corrupt. Translation in brackets follows Lu's *Corrigenda* 2b. Cf. Legge, II, p. 411.

he will look up and down, and hesitate between advancing and retreating."

5.18 (7a1) Someone said: "Can you [discuss the above] with specific reference to the Confucian Canons?"

I said: "The *I-ching* states, 'The way of the *Ch'ien* (the first hexagram) works through change and transformation; thus each thing is endowed with its appropriate nature and destiny.'[66] This means that the myriad things each have their natures. It also says: '[*Hsien* (the 31st hexagram) means stimulation (or influence).... As Heaven and Earth stimulate each other, all things take shape and come into being. When the Sage stimulates the hearts of men, the world is at peace.] If we examine this stimulating process, we can see the feelings[67] of Heaven and Earth and the myriad beings.' This means that feelings are a 'process of responses' to stimuli. Since even the insects, the grass, and the woods have their natures, natures cannot be all good (in a moral sense). Since Heaven, Earth, and the Sage are said [to possess stimulated] feelings, feelings cannot be all bad.

"It also says: '[The meaning of the eight hexagrams is expressed through symbols]. The explanation of the lines [of a hexagram] (*hsiao-tz'u*) and the judgment about the hexagram itself (*t'uan*) communicate the feelings [of the hexagram].'[68] This refers to the same concept [of feelings discussed above].

"The terms *ch'ing* (feeling), *i* (idea), *hsin* (the heart), and *chih* (ideal) are but different designations for *hsing* (human

---

[66] *I-ching* 1:6a, Legge, *SB*, XVI, p. 213; R. Wilhelm, *I-ching*, II, p. 4.

[67] *Ch'ing*, see note 57. Since the citation of *I-ching* 4:1b–2a refers specifically to the word *ch'ing*, Legge's rendering of it as "character" and Wilhelm's as "nature" both miss the point. Here the emphasis is on the stimulating process and its relation with *ch'ing* (feeling). A full citation from the *I-ching* describing the stimulating influences is inserted here in brackets. See Legge, *SB*, XVI, p. 238; R. Wilhelm, *I-ching*, II, pp. 184–185.

[68] *I-ching* 8:24a, Legge, *SB*, XVI, p. 405; R. Wilhelm, *I-ching*, I, pp. 380–381. Here again both Legge's and R. Wilhelm's rendering of the character *ch'ing* miss the point.

nature) in action. [Thus the *I-ching*] states: 'The feelings [of the Sages] are evident in their explanations [of the *I-ching*].'[69] This refers to feelings, *ch'ing*. [It also] states: 'Speech cannot fully express ideas.'[70] This refers to ideas, *i*. [The ode reads]: 'In my heart, I love him.'[71] This refers to the heart, *hsin*. It is also said that a man should exercise control [over what he adopts as] his ideals.[72] This refers to ideals, *chih*. These are merely different names appropriate for different occasions. Why must only *ch'ing* (feelings) be [regarded as] responsible for what is evil? Thus it is said: 'It is necessary to rectify names.'"[73]

5.19 (7b5) Someone said: "If good and evil are both [constituents of] human nature, then what is the use of law and education?"

I said: "Although human nature has good [inclinations], they need to be perfected by education; although human nature has bad [inclinations], they can be mitigated by law. Only the highest Sage and the lowest moron cannot be changed.[74] All others experience an [inner] conflict between the good and bad [inclinations of their natures]; whence education enhances the good and law suppresses the bad.

"Of the nine types of people,[75] one half of them may be

---

[69] *I-ching* 8:3b, Legge, *SB*, XVI, p. 381; R. Wilhelm, *I-ching*, I, p. 352.

[70] *I-ching* 7:30b, Legge, *SB*, XVI, p. 377.

[71] Legge, IV, p. 279.

[72] From *Tso-chuan*, Legge, V, p. 708, "[The Sage-kings] carefully followed these relations and analogies [in formulating rites by which] to regulate those six impulses, *chih*." For "rectification of names," see p. 39.

[73] Legge, I, p. 263. Hsün Yüeh apparently produced these quotations from the Confucian canons in order to support his high estimation of the value of *ch'ing*. They convey the idea that the Sages expressed feelings, while speech cannot fully express ideas; feelings (love) come directly from the heart, while ideals need to be regulated.

[74] Legge, I, p. 318.

[75] *Chiu-p'in*, nine grades. The classification of personalities into nine grades (the upper, middle, and lower, each with three subdivisions—upper, middle, and lower) was first utilized by Pan Ku in *HS* 20. In A.D. 220 this became an

responsive to educational guidance and three-fourths of them may be constrained by law. Those who cannot be changed may amount to roughly one-ninth [of the people]; but, of this one-ninth, there may still be some who can be influenced slightly. Accordingly, law and education may have a reforming effect upon almost all people. On the other hand, when they (law and education) are misused, they will have a detrimental effect as great [as their beneficial impact would have been]."

5.20 (8a3) Someone said: "When law and education are properly tended to, the state will be in good order; when they are not, there will be disorder. But if they are neither properly tended to nor disregarded and the people are allowed to give free rein to their passions,[76] will the state find itself in a condition between good and bad?"

I said: "That which is *yang* always ascends; that which is *yin* always descends.[77] Ascent is difficult; descent is easier. Goodness is *yang* and badness is *yin*. Therefore, it is difficult to be good and easy to be bad. If people's passions are allowed free rein, there will be more [people] falling into a lower state [of moral consciousness than those rising to a higher one]."

Someone said: "Then where are the middle ones?"

I said: "When law and education are not pure, [that is, when they are] partly good and partly bad, the state will find itself in a condition between good and bad."

5.21 (8a8) Best is he whose virtue is pure and flawless; next best is [he whose evil nature is] dormant and inert; next

---

established system of personality evaluation in the selection and promotion of officials under the Wei dynasty. See Miyazaki Ichisada, *Kubon Kanjin ho no kenkyu* (Kyoto, 1956), pp. 8–12, 92–105.

[76] *Ch'ing*, see note 57.

[77] For Hsün Shuang's exposition of the *"sheng-chiang"* (rise-and-decline) theory to interpret the *I-ching*, see my article, "A Confucian Magnate's Idea of Political Violence: Hsün Shuang's (128–190) Interpretation of the *Book of Changes*," *T'oung-pao* 54 (1968), 73–115.

THE SHEN-CHIEN

is [he whose evil nature] is active [in terms of producing evil thoughts], but not yet evil deeds, [he whose evil thoughts have] given rise to evil-doing that has not yet gone far, or [he whose evil-doing] has gone far, but is still able to be reversed.[78] Worst is [he whose evil-doing] has gone too far [to be turned around].[79]

All these are [expressions of] man's nature. It is the mind which controls them. He who is able to suppress [his evil nature] when it has become active or is able to stop [his evil-doing] when it has been set in motion has the same nature as a person of higher caliber. He who does not stop his [evil-] doing or goes too far [for a reversal][80] ends up the same as a person of lower caliber.

5.22 (8b5) The superior man prizes benevolence, but does not demand favors; he honors propriety of conduct, but does not question motives;[81] he exalts virtue, but does not reprimand[82] those who air grievances. When he demands or reprimands, he does so [with reference] to himself first. And when he does a good deed, he also does it before anyone else.[83] [Demanding or giving] too many favors, bending one's motives,[84] and harboring secret grievances—these three things are truly detrimental to the Way of moral integrity and injurious to [the cultivation of] great virtue. They are decisive in [determining] one's successes and failures, gains and losses. If everyone is desperately [busy] remedying the ills [of the

---

[78] *Fu*, recover, reverse, conversion, also the designation of the 24th hexagram in the *I-ching*. See Legge, *SB*, xvi, pp. 107–108.

[79] Lit., "far and not near." *CSCY* 46:9b reads "遠而已矣 nothing but far." My translation follows the context of the preceding clause.

[80] See note 79.

[81] *I*, ideas, thoughts; see 5.18 above. The term here refers to something inward in contrast to the propriety of outward conduct (*li*). The emphasis on conduct rather than motive comes close to Legalism. See note 57.

[82] Note the intricate meaning of the character 責 *tse*, rendered variably as to demand, to question, and to reprimand in these three clauses.

[83] *Hsien-jen* may mean either to do it before others or to give priority to others.

[84] One who anticipates the intention of others in order to please.

THE SHEN-CHIEN

world], who will have time to care about the Way and [the cultivation of] moral virtue? Such is the condition of a corrupt age.[85]

5.23 (8b9) Superior men are eternally bonded together[86] [through their commitment to] righteousness, and eternally pledged to each other by their good faith.[87] How narrow it is [for people to feel] they must socialize before they can become intimate [friends], and must take oaths [to ensure] their friendship![88]

Greatest is he who does not discriminate between the past and the present; next best is he who does not discriminate between the regions bounded by the seas.[89] What a wonderful virtue it is for one's ideal to be in union with all under Heaven![90]

The ideal of a great man is intangible; it is immense[91] and in union[92] with the Way. The ideals of the common people are obvious; they clearly conform with custom. When an ideal is in union with the Way, it does not float and sink with [the vicissitudes of] vulgar custom.

5.24 (9a7) Someone said: "Is cultivating one's conduct not [to please] other [people] but out of [a feeling of] shame before the Divine Intelligence the highest [degree of excellence]?"[93]

---

[85] Note the outspoken pessimism in this passage. Such a dim view of human nature often served as a justification for Legalist practices.

[86] *Chiao*, friendship, social intercourse, meeting, communion.

[87] They rely on righteousness and good faith rather than vulgar social contact and oath to cement their unity.

[88] 故 *ku* (old [acquaintance], friendship) reads 固 *ku* (permanent, to ensure) in edition no. 4.

[89] The entire country of China.

[90] It may also read "unifying all under Heaven."

[91] *Hao-jan*, all-pervasive. See Legge, II, pp. 189–190.

[92] *T'ung*, identical with. Compare this with the notion of "reach" (*chih*) in note 11.

[93] For the importance of the feeling of shame in Confucian teachings, see Legge, I, pp. 143, 146, 182, 275, 407; II, pp. 451–452. Divine Intelligence, *shen-ming*; see notes 2 and 3. Regarding the superlative *chih*, see note 11.

I said: "Not quite. To feel shame in one's [own conscience] is primary; to feel shame before the Divine Intelligence is second best; and to feel shame before other people is [to be concerned with] external things. [If a man is concerned only with] that which is external, he will accumulate evil inside. Hence the superior man will carefully examine [whether he is capable of feeling] shame in his own [conscience]."[94]

5.25 (9b2) Someone said: "Is [having a sense of] shame itself the ultimate ideal?"[95]

I said: "Not quite. A man's ideal[96] springs naturally from within him. What does it have to do with shame?"

[If a man] runs toward a ravine, he will fall into it; if he loses [control] in the water, he will drown. People are aware of this. If a man runs toward a trap, he will fall into it; if he loses the Way, he will sink [into evil]. But people are not aware of such things—they overlook them. Therefore the superior man is careful of what he might overlook.[97]

If a man does not listen to the highest counsel, his ideals cannot be expanded. If he does not listen to the highest teachings, his mind cannot become firm. When a man recalls [the Sage-rulers of] T'ang and Yü[98] of remote antiquity and looks up to Confucius in medieval times,[99] then he knows that [he who follows] an inferior Way is contemptible. When a man thinks about Po-i at Mount Shou-yang[100] and remembers the

---

[94] For the emphasis on feeling shame in one's own conscience in Confucian teachings, see Legge, I, pp. 366–367. See also Leo Wieger, *A History of the Religious Beliefs and Philosophical Opinions in China*, tr. by E. C. Werner (New York, 1969 reprint), pp. 341.

[95] 志 *chih* (determination, ambition, ideal) reads 至 *chih* (to reach, ultimate) in *Fu-tzu* (Wu-ying tien chü-chen pan ed.) 2b. This passage is entirely identical to the "*jen-lun*" (Discourse on Kindheartedness) section in *Fu-tzu* 1:2b8–3b2.

[96] *Chih* reads 至 *chih* in *Fu-tzu*; see notes 11 and 95.

[97] Legge, I, p. 384.

[98] Kingdoms of the legendary Sage-kings Yao and Shun of antiquity.

[99] *Chung-ku* (medieval or the recent past) as compared with *shang-ku* (remote antiquity).

[100] Po-i, see *SC* 1, note 75.

Four Elders of Mount Shang,[101] then he knows that [he who cherishes] an impure ideal is shameful. When a man keeps in mind Chang Ch'ien's [adventure] in the far Western Regions[102] and ponders Su Wu's [mission] to the northern borderland,[103] then he knows that he who feels attached to his native place and to his family is disgraceful. Reasoning from such examples,[104] there is no excellence one cannot attain. Compare your virtue with those above you and your desires with those below you. By comparing your virtue with those above you, you will know shame; by comparing your desires with those below you, you will know contentment. When you know shame, then you are almost a Sage. When you know contentment, then you can be at peace in any unfavorable situation. If you are almost a Sage, how can there be any evil [in your mind?] If you can be at peace in any unfavorable situation, how can you ever go to extremes? This is called having a standard of discipline.[105]

The purest of the pure is the highest; the second best only approximates this [ideal].[106] There is now no one [who does better than] to approximate [the ideal]. If it can even be approximated, there will be no evil; this is good enough. The superior man engages in daily self-reflection. When angry he does not violate [the commands of] virtue; when happy he does not violate [the precepts of] righteousness.[107]

---

[101] Four Elders of Mt. Shang; see *SC* 4, note 39.

[102] Chang Ch'ien, the diplomat who explored the Western Regions; biography in *HS* 61:1-5a. See also Friedrich Hirth, "The Story of Chang K'ien, China's Pioneer in Western Asia," *JAOS* 37 (1919), 89-152.

[103] See *SC* 4, note 50.

[104] *T'ui*, to reason by inference from analogies or examples.

[105] *Chien*, see *SC* 4, note 8.

[106] *Kai*, general, approximate. Note the pessimistic tone here. See also the discussion of *chih* in note 11.

[107] According to Lu's *Corrigenda* 2b, *ssu* (four) should read 日 *jih* (day). The character reads 内 *nei* (inner) in *Fu-tzu* 1:3b. There is another omission in the *SPTK* edition, which according to *Fu-tzu* should read 亂 *luan* (violate).

BIBLIOGRAPHY

Balazs, Etienne, *Chinese Civilization and Bureaucracy*, tr. by H. M. Wright and ed. by Arthur F. Wright, Yale University Press, 1964.

Bauer, Wolfgang, *China and the Search for Happiness*, tr. by Michael Shaw, New York, 1976.

Bielenstein, Hans, "An Interpretation of the Portents in the *Ts'ien Han shu*," *Bulletin of the Museum of Far Eastern Antiquity*, 22(1950), 127–143.

"The Restoration of the Han Dynasty (I)," *Bulletin of the Museum of Far Eastern Antiquity*, 26 (1954), 5–210.

Bodde, Derk, *China's First Unifier: A Study of the Ch'in Dynasty as Seen in the Life of Li Ssu*, Leiden, 1938.

CC *Hou-Han-shu chi-chieh*.

CSCY *Ch'ün-shu chih-yao*.

Chan-kuo ts'e 戰國策, *Ssu-pu ts'ung-k'an* edition.

Chan, Wing-tsit, tr. and compiled, *A Source Book in Chinese Philosophy*, Princeton University Press, 1963.

Chang Hsin-cheng, 張心澂, *Wei-shu t'ung-k'ao* 僞書通考, Shanghai, 1939, 1954 reprint.

Chang, K. C., *Early Chinese Civilization*, Harvard University Press, 1976.

Chang, Tung-sun, "A Chinese Philosopher's Theory of Knowledge," tr. by Li An-che, *The Yenching Journal of Social Studies*, 1:2 (January 1939), 161–195.

Chao I 趙翼, *Nien-erh shih cha-chi* 廿二史箚記, Shang-wu yin-shu kuan, 1958.

Chavannes, Edouard, tr., *Les Mémoires Historiques de Se-ma Ts'ien*, 5 vols., Paris, 1895–1905.

Chen, Chi-yun, "A Confucian Magnate's Idea of Political Violence: Hsün Shuang's (A.D. 128–190) Interpretation of the Book of Changes," *T'oung-pao* 54 (1968), 73–115.

Hsün Yüeh (A.D. 148–209): The Life and Reflections of an Early Medieval Confucian, Cambridge University Press, 1975.

Chen-kuan cheng-yao 貞觀政要, *Ssu-pu ts'ung-k'an* edition.

BIBLIOGRAPHY

Chih-chai shu-lu chieh-t'i 直齋書錄解題, by Ch'en Chen-sun 陳振孫, Kuang-ya shu-chü edition.

Chin-shu 晉書, *Po-na* edition.

Chou Hung-hsiang 周鴻翔, *Pu-tz'u tui-chen shu-li* 卜辭對貞述例, Hong Kong, 1969.

Chou-i cheng-i 周易正義, in *Shih-san ching chu-shu*, I-wen yin-shu-kuan photolithic edition.

Chou-li see *Chou-li chu-shu*.

Chou-li chu-shu 周禮注疏, 1892 reprint of the 1815 edition.

Chou-shu 周書, *Po-na* edition.

Chuang-tzu 莊子, *Ssu-pu pei-yao* edition.

Chün-chai tu-shu chih 郡齋讀書志, by Chao Kung-wu 晁公武; *Fu-chih* 附志, by Chao Hsi-pien 趙希弁; Commercial Press, 1937 reprint of the 1249, 1250, and 1511 editions.

Chung-hsing kuan-ko shu-mu 中興館閣書目, fragmentary in *Ku-i-shu-lu ts'ung-chi*, Pei-p'ing, 1933.

Ch'en Ch'i-yün see Chen, Chi-yun.

Ch'en Meng-chia 陳夢家, *Yin-hsü pu-tz'u tsung-shu* 殷墟卜辭綜述, Peking, 1956; Tokyo, 1964 reprint.

Ch'eng, Chung-ying, "Inquiries into Classical Chinese Logic," *Philosophy East and West* 15 (July-August 1965), 195–216.

Ch'eng, Chung-ying and Richard S. Swain, "Logic and Ontology in the *Chih-wu lun* of Kung-sun Lung Tzu," *Philosophy East and West* 20 (April 1970), 137–154.

Ch'ien Mu 錢穆, *Ch'in Han shih* 秦漢史, Hong Kong, 1957.

Ch'u-hsüeh chi 初學記, Huang-shih 1888 edition.

Ch'ü, T'ung-tsu, *Law and Society in Traditional China*, Paris, 1961.

Ch'un-ch'iu Ku-liang chu-shu 春秋穀梁注疏, I-wen yin-shu kuan reprint of the 1815 edition.

Ch'un-ch'iu Kung-yang chu-shu 春秋公羊注疏, I-wen yin-shu kuan reprint of the 1815 edition.

Ch'ün-shu chih-yao 羣書治要, *Ssu-pu ts'ung-k'an* edition.

Ch'ün-shu shih-pu 羣書拾補, *Pao-ching t'ang ts'ung-shu* edition.

Ch'ung-wen tsung-mu 崇文總目, *Yüeh-ya t'ang ts'ung-shu* edition.

Creel, Herrlee G., *Chinese Thought From Confucius to Mao Tse-tung*, Mentor Book, 1953.

    Confucius and the Chinese Way, Harper Torchbook, 1960

reprint of *Confucius the Man and the Myth*, 1949.
What is Taoism?, University of Chicago Press, 1970.
The Origins of Statecraft in China, University of Chicago Press, 1970.
Crump, J. I., Jr., *Intrigue: Studies of the Chan-kuo ts'e*, University of Michigan Press, 1964.
Dawson, Raymond, ed., *The Legacy of China*, Oxford University Press, 1964.
De Bary, Wm. Theodore, et al, ed., *Sources of Chinese Tradition*, Columbia University Press, 1960.
Dubs, H. H., tr., *The Works of Husntze*, London, 1928.
Dubs, I-III Dubs, H. H., tr., *The History of the Former Han Dynasty*, 3 vols., Baltimore, 1938, 1944 and 1955.
Duyvendak, J. J. L., tr., *Tao Te Ching*, London, 1954.
Fairbank, John K., ed., *Chinese Thought and Institutions*, University of Chicago Press, 1957.
Fork, Alfred, tr., *Lun-heng*, New York, 1962 reprint.
Fu Ssu-nien 傅斯年, *Hsing-ming ku hsün pien-cheng* 性命古訓辨証, Shang-wu yin-shu kuan, 1940.
Fu-tzu 傅子, Wu-ying-tien chü-chen pan edition.
Fung, Yu-lan, *A History of Chinese Philosophy*, tr. by Derk Bodde, 2 vols., Princeton University Press, 1952-1953.
A Short History of Chinese Philosophy, ed. by Derk Bodde, New York, 1960.
Gardner, Charles S., *Chinese Traditional Historiography*, Harvard University Press, 1938, 1961.
Graham, A. C., "Kung-sun Lung's Essay on Meaning and Things," *Journal of Oriental Studies* 2:2 (1955), 282-301.
tr., *The Book of Lieh-tzu*, London, 1960.
"Two Dialogues in Kung-sun Lung Tzu: 'White Horse' and 'Left and Right,'" *Asia Major* (N. S.) 11:2 (1965), 128-152.
Granet, Marcel, *La Pensee Chinoise*, Paris, 1934.
HC Han-chi.
HHC Hou-Han chi.
HHS Hou-Han shu.
HJAS Harvard Journal of Asiatic Studies.

BIBLIOGRAPHY

HS Han-shu.
HSPC Han-shu pu-chu.
HWTS Han-Wei ts'ung-shu.
Han-chi 漢紀, by Hsün Yüeh 荀悅, *Ssu-pu ts'ung-k'an* edition.
Han Fei tzu 韓非子, *Ssu-pu pei-yao* edition.
Han-shih wai-chuan 韓詩外傳, *Han-Wei ts'ung-shu* edition.
Han-shu 漢書, *Po-na* edition.
Han-shu pu-chu 漢書補注, by Wang Hsien-ch'ien 王先謙, Wang-shih 1915 edition.
Han, Yü-shan, *Elements of Chinese Historiography*, Hollywood, California, 1955.
Hansen, Chad D., "Ancient Chinese Theories of Language," *Journal of Chinese Philosophy* 2:3 (June 1975), 245–283.
Hightower, J. R., tr., *Han-shih wai-chuan*, Harvard University Press, 1952.
Holzman, Donald, *Poetry and Politics: The Life and Works of Juan Chi, A.D. 210–263*, Cambridge University Press, 1976.
Hou-Han i-wen-chih 後漢藝文志, by Yao Chen-chung 姚振中, in *Erh-shih wu shih pu-pien*, K'ai-ming shu-tien edition.
Hou-Han chi 後漢紀, by Yuan Hung 袁宏, Commercial Press: *Wan-yu wen-k'u* edition.
Hou-Han shu 後漢書, *Po-na* edition.
Hou-Han shu chi-chieh 後漢書集解, by Wang Hsien-ch'ien 王先謙, Wang-shih 1915 edition.
Huai-nan tzu 淮南子, *Ssu-pu ts'ung-k'an* edition.
Hui Tung 惠棟, *I Han-hsüeh* 易漢學, in *Huang-Ch'ing ching-chieh, hsü-pien*.
I-li 易例, in *Huang-Ch'ing ching-chieh, hsü-pien*.
Hulsewé, A. F. P., *Remants of Han Law*, I, Leiden, 1955.
Hsin T'ang-shu 新唐書, *Po-na* edition.
Hsü, Cho-yün, *Ancient China in Transition: An Analysis of Social Mobility, 722–222 B.C.*, Stanford University Press, 1965.
Hsün-tzu 荀子, *Ssu-pu pei-yao* edition.
  Also *Ssu-pu ts'ung-k'an* edition.
Hsün Yüeh see Chen, Chi-yun.
I-ching Chou-i chu-shu.

BIBLIOGRAPHY

I-lin 意林, *Ssu-pu ts'ung-k'an* edition.
I-wen lei-chü 藝文類聚, Wang-shih 1589 edition.
JAOS Journal of the American Oriental Society.
Jan, Yün-hua, "The Silk Manuscripts on Taoism," *T'oung-pao* 63:1 (1977), 65–84.
Kaltenmark, Max, tr., *Le Lie-sien tchouan*, Peking, 1953.
Ku Yen-wu 顧炎武, *Jih-chih lu* 日知錄, Commercial Press: *Wan-yu wen-k'u* edition.
Kuan Feng and Lin Lü-shih, "Development of Thought and the Birth of Materialist Philosophy at the End of the Western Chou and the Beginning of the Eastern Chou," *Chinese Studies in Philosophy* 2:1–2 (1970–71), 54–79.
"Thought of the Yin Dynasty and the Western Chou," *Chinese Studies in Philosophy* 2:1–2 (1970–71), 4–53.
Kuan-tzu 管子, *Ssu-pu ts'ung-k'an* edition.
Kung-sun Lung tzu 公孫龍子, *Ssu-pu pei-yao* edition.
Kuo-yü 國語, *Ssu-pu ts'ung-k'an* edition.
Lao, Kan 勞榦, "Kuan-yu Han-tai kuan-feng ti chi-ko t'ui-ts'e" 關於漢代官俸的幾個推測, *Wen-shih-che hsüeh-pao* 3 (1951), 11–22.
Legge, I-v Legge, James, tr., *The Chinese Classics*, Vols. 1–5.
Legge, James, tr., *The Chinese Classics*, 5 vols., Hong Kong University Press, 1960 reprint.
The Yi-king, in F. Max Muller ed., *The Sacred Books of the East*, Vol. 16, Oxford, 1899.
Li-chi Li-chi chu-shu.
Li-chi chu-shu 禮記注疏, I-wen yin-shu kuan reprint of the 1815 edition.
Libenthal, Walter, tr., *The Book of Chao*, Peking, 1948.
Lieh-nü chuan 列女傳, *Ssu-pu ts'ung-k'an* edition.
Lieh-hsien chuan 列仙傳, *Ts'ung-shu chi-ch'eng* edition.
Lieh-tzu 列子, *Han-Wei ts'ung-shu*, fourth series, 1880.
Loewe, Michael, *Crisis and Conflict in Han China*, London, 1974.
"Manuscripts Found Recently in China: A Preliminary Survey", *T'oung-pao* 68:2–3 (1977), 99–136.
Lun-yü ku-chu chi-chien 論語古注集箋, Ch'iang-su shu-chü, 1881 edition.

BIBLIOGRAPHY

Maspero, Henri, *Mélanges posthumes sur les religions et l'histoire de la Chine*, Paris, 1950.

Mather, Richard B., tr., *Shih-shuo hsin-yü: A New Account of Tales of the World*, University of Minnesota Press, 1976.

Metzger, Thomas A., *Escape from Predicament: Neo-Confucianism and China's Evolving Political Culture*, Columbia University Press, 1977.

Miyazaki, Ichisada 宮崎市定, *Kubon kanjin hō no kenyu* 九品官人法の研究, Kyoto, 1956.

Mote, Frederick W., *Intellectual Foundations of China*, New York, 1971.

Munro, Donald J., *The Concept of Man in Early China*, Stanford University Press, 1969.

Nan-ch'ing cha-chi 南菁札記, 1894 edition.

Needham, Joseph, *Science and Civilization in China*, Cambridge University Press, Vol. 2, 1956; Vol. 3, 1959.

Nei-ko ts'ang-shu mu-lu 內閣藏書目錄, in *Shih-yuan ts'ung-shu*.

Nivison, David, and Arthur F. Wright, ed., *Confucianism in Action*, Stanford University Press, 1959.

O'Hara, Albert R., *The Position of Women in Early China*, The Catholic University of America Press, 1945.

PC Han-shu pu-chu.

PTSC Pei-t'ang shu-ch'ao.

Pao-p'u tzu 抱朴子, *Tao-tsang* edition.

Pei-t'ang shu-ch'ao 北堂書鈔, Nan-hai K'ung-shih 1888 edition.

Pokora, Timotheous, tr., *Hsin-lun (New Treatise) and Other Writings by Huan T'an (43 B.C.–A.D. 28)*, Michigan Papers in Chinese Studies 20, 1975.

P'i Hsi-jui 皮錫瑞, *Ching-hsüeh li-shih* 經學歷史, Peking, 1959.

Rickett, W. Allyn, tr., *Kuan-tzu*, Hong Kong University Press, 1965.

Rosemont, Henry, Jr., "On Representing Abstractions in Archaic Chinese," *Philosophy East and West* 24:1 (January 1970), 71–88.

——— "State and Society in *Hsün-tzu*: A Philosophical Commentary," *Monumenta Serica* 29 (1970–1971), 38–78.

SB The Sacred Books of the East.

BIBLIOGRAPHY

SC *Shen-chien.*
SKC *San-kuo chih.*
SPPY *Ssu-pu pei-yao.*
SPTK *Ssu-pu ts'ung-k'an.*
The Sacred Books of the East, ed. by E. Max Muller, Oxford, 1880–1910.
San-kuo chih 三國志, *Po-na* edition.
San-kuo chih chi-chieh 三國志集解, by Lu Pi 盧弼, I-wen yin-shu kuan reprint edition.
Schafer, Edward H., "Ritual Exposure in Ancient China," *Harvard Journal of Asiatic Studies* 14 (1951), 130–144.
Seikado hiseki-shi 靜嘉堂祕籍志, Tokyo, 1917–1919.
Solomon, Bernard S., "The Assumptions of Hui-tzu," *Monumenta Serica* 28:1 (1969), 1–40.
Sui-ch'u t'ang shu-mu 遂初堂書目, by Yu Mou 尤袤, in *Shuo-fu*, Han-fen lou edition.
Sui-shu 隋書, *Po-na* edition.
Sui-shu ching-chi-chih k'ao-cheng 隋書經籍志考証 by Yao Chen-chung 姚振宗, in *Erh-shih wu shih pu-pien*.
Sun, Etu Zen, ed., *Chinese Social History*, Washington, D.C., 1956.
Swann, Nancy Lee, *Pan Chao: Foremost Woman Scholar of China*, New York, 1932.
    Food and Money in Ancient China, Princeton University Press, 1950.
Shan-hai ching 山海經, *Ssu-pu ts'ung-k'an* edition.
Shang-shan t'ang shu-mu 上善堂書目, Ch'en-shih 1929 edition.
Shen-chien 申鑑, by Hsün Yüeh 荀悅, *Ssu-pu ts'ung-k'an* edition.
Shih-chi 史記, *Po-na* edition.
Shiki *Shiki kaichū kōshō.*
Shiki kaichu kōshō 史記會注考証, Tokyo, 1934.
Shih-t'ung 史通, by Liu Chih-chi 劉知幾, *Ssu-pu ts'ung-k'an* edition.
Shui-ching chu 水經注, Wu-yin tien chü-chen pan edition.
Shuo-wen chieh-tzu ku-lin 說文解字詁林, Taiwan: Lithographic reprint of the 1931 edition.
Shuo-yuan 說苑, *Han-Wei ts'ung-shu* edition.

BIBLIOGRAPHY

Ssu-k'u ch'üan-shu tsung-mu, t'i-yao 四庫全書總目提要, Commercial Press: *Wan-yu wen-k'u* edition.

Ti-fan 帝範, *Kuang-ya shu-chü* edition.

Tjan, Tjoe Som, *Po Hu T'ung*, Leiden, 1949.

Trauzettel, Von Rolf, "Zum Problem der chinesischen Ontologie unter dem Aspekt der Sprache," *Zeitschrin der Deutschen Morgenländischen Gesellschaft* 119:2 (1970), 270–277.

T'ai-p'ing yü-lan 太平御覽, Taiwan, Hsin-hsing shu-chü reprint of the Kanezawa bunko edition.

T'ang-shu 唐書, *Po-na* edition.

T'ieh-ch'in t'ung-chien lou ts'ang-shu mu-lu 鐵琴銅劍樓藏書目錄, Ch'ü-shih edition.

T'ien-lu lin-lang shu-mu 天祿琳琅書目, Wang-shih 1884 edition.

T'ung-chih 通志, by Cheng Ch'iao 鄭樵, Commercial Press: *Wan-yu wen-k'u* edition.

Tseng-ting Ssu-k'u chien-ming mu-lu 增訂四庫簡明目錄, Chung-hua shu-chü, 1959.

Ts'ang-yuan ch'ün-shu t'i-chi 藏園羣書題記, 1938.

Tso-an chi 左盦集, Liu Shen-shu hsien-sheng i-shu edition.

Ts'ung-shu chü-yao 叢書舉要 by Yang Shou-ching 楊守敬, I-ch'iu-kuan edition.

Tzu-chih t'ung-chien 資治通鑑, by Ssu-ma Kuang 司馬光, Chung-hua shu-chü, 1956.

Tz'u-ch'i Huang-shih jih-ch'ao fen-lei 慈溪黃氏日抄分類 by Huang Chen 黃震, Tz'u-ch'i Feng-shih keng-yü lou edition.

Uno, Seiichi, "Some Observations on Ancient Chinese Logic," *Philosophical Studies of Japan* 6 (1965), 31–42.

Waley, Arthur, tr., *The Way and Its Power*, New York, 1956.

Wang Ch'ung 王充, *Lun-heng* 論衡, *Han-Wei ts'ung-shu*, First Series.

Wang Fu 王符, *Ch'ien-fu lun* 潛夫論, *Han-Wei ts'ung-shu*, First Series.

Wang, Yü-ch'üan, "An Outline of the Central Government of the Former Han Dynasty," *Harvard Journal of Asiatic Studies* 12 (1949), 134–187.

Watson, I-II Watson, Burton, tr., *Records of the Grand Historian of China*, Vols. I-II, Columbia University Press, 1961.

Watson, Burton, *Ssu-ma Ch'ien: Grand Historian of China*, Columbia University Press, 1958.

Wen-hsien t'ung-k'ao 文獻通考, by Ma Tuan-lin 馬端臨, Commercial Press: *Wan-yu wen-k'u* edition.

Wen-hsüan 文選, *Ssu-pu ts'ung-k'an* edition.

Wen-yüan-ko shu-mu 文淵閣書目, *Tu-hua chai ts'ung-shu* edition.

Wieger, Leo, *A History of the Religious Beliefs and Philosophical Opinions in China*, tr. by E.C. Werner, New York, 1969 reprint.

Wilhelm, Hellmut, *Change: Eight Lectures on the I-ching*, tr. by Cary F. Baynes, London, 1960.

Wilhelm, Richard, tr., *I-ching*, English translation by Cary F. Baynes, New York, 1950.

Wright, Arthur F., *Buddhism in Chinese History*, Stanford University Press, 1959.

Wu, K.T., "Ming Printing and Printers," *Harvard Journal of Asiatic Studies* 7 (1943), 203-260.

Wu-shih-wan chüan lou ch'ün-shu pa-wen 五十萬卷樓羣書跋文, 1949 edition.

Wu-shih-wan chüan lou ts'ang-shu mu-lu 五十萬卷樓藏書目錄, 1936 edition.

Wu-Yüeh ch'un-ch'iu 吳越春秋, *Han-Wei ts'ung-shu*, Second Series.

Yang Hsiung 揚雄, *Fa-yen* 法言, *Han-Wei ts'ung-shu*, First Series.

Yang, Lien-sheng, *Money and Credit in China*, Harvard University Press, 1952.

Studies in Chinese Institutional History. Harvard University Press, 1963

Yen-tzu ch'un-ch'iu 晏子春秋, *Ssu-pu ts'ung-k'an* edition.

Yü-hai 玉海, 1687-1688 edition.

Zürcher, Erik, *The Buddhist Conquest of China*, Leiden, 1959.

INDEX-GLOSSARY

abdication of throne, 24–25, 26
abilities, 45. *See also ts'ai*
absolute, attainable by man, 8–9; uncertainty of, 8
absolute proprietary right, 139n
absolutism, 53
action, 46, 179, 180n. *See also hsing* 行, conduct.
acupuncture, 160, 165
administration of the harem, 144
administrators of the commanderies, 130
Age of Disunity, 52, 53, 92
Age of Great Unity, *see ta-t'ung*
Agnosticism, 5–9, 53, 91–92, 150–51, 161, 182–83
agriculture, 108, 109n
ai 愛, 157
alchemism, 160
ambivalence, 44, 53, 104–105n, 125n
amnesty, 127, 146–47; in Han times, 116n; mentioned in *Chou-li*, 116n
Ancestor P'eng 彭祖, 156
ancestors, *see tsu* and *tsung*
ancestral temples, 148
anthropocentric optimism, 12n
anthropocentric world-view, 8, 13, 36, 106n; the universe likened to man's abode, 152
ao 奧, 185
apocryphal and prognostic traditions and works, 27–28, 43. *See also ch'an* and *wei* 緯
appointment of court official to regional and local government, 130–31
approximation, 51, 197
arrest and blacklisting of partisans, *see tang-ku*
art of war, 132

astrology, 143
autonomy versus heteronomy, 30n
average person, 193. *See also chung-jen*, ordinary human being

balance, 160; in one's deportment, 179
bandits and outlaws, 131
barbarians, 131, 149, 175
benevolence and righteousness, *see jen-i*
biography of Hsün Yüeh, 74, 75
blood vengeance, 126, 136–37
Book of Changes, see *I-ching*
Book of Historical Documents, see *Shu-ching*
breathing exercise, 159n. *See also* magic and Taoism
bribery, 110
Buddhism, 4
Buddhist, 52
business transaction, 141

calendar, 152
central government, 133n, 134n
change according to times, 86, 95, 127, 132–33, 134–36, 138
change of dynasties, 24–25, 27n
choice, 47, 167, 188n; moral, 190
cinnabar, 160n
circulation of goods, 140
civil authority, 111
civil virtue, 97
civil war, 33
civilization, the continuity of, 11
classical learning, 14n
Classics, 99, 105n, 145. *See also* Confucian canons, *ching* 經
coercion, 112, 120

209

commerce, 109n, 140–41; tax of, 122n
commoners, 113, 114, 118, 163
communication, 185
complimentary interaction of Heaven and Earth, 164
Comprehensive Mirror for Aid in Government, see *Tzu-chih t'ung-chien*
comprehensiveness, see *t'ung* 通
compromise, 116n. See also gradualism
conduct, 130, 178–82, 194–95. See also hsing 行
conflict, between the centrifugal and centripetal forces during the Later Han, 32; between "Heavenly reason" and "human desire or emotion", 47n; between innate goodness and harmful environment, 47n; between the "inner" and the "outer" worlds, 30–31; between man's innate badness and his desire for goodness, 47n
conformity, 113, 114, 195
Confucian Canons, 74, 185, 191, 192n; Erudites, 27n; literati, 27n, 30n
Confucianism, 16n, 35n, 52, 96, 108, 179n, 180n, 181n, 188n; ambivalence toward the state, 21; Ancient-text School, 134n, 145; apathy toward the ruling dynasty, 26; attitude of the Former Han rulers toward, 96; common thesis, 51; conformity, 8n; controversies within, 39–40, 145–46; discredited, 4; disparate dates given for its texts, 145; eclectic, 7, 8, 14, 17; Emperor Hsüan's comment on, 72; emphasis on the feeling of shame, 195–96n; idea of the relationship between Heaven and man, 19–20; ideal of land system, 95n; ideal of political unity, 94n; ideas about human nature, 45, 187; Kung-yang and Ku-liang Schools, 94n; Modern-text School, 134n, 145; moral idealism, 4; moral integrity, 26; orthodox moralism, 27; orthodoxy, 4; orthodoxy of the imperium, 23; polarity between *li* and *ch'i*, *hsing* and *ch'ing*, the good and the bad, 188n; principle of minimal legal restraint, 135n; pristine spirit of, 5, 48; reforms, 23; rituals and music, 153n; use of moral suasion, 135n; utopia, 139n. See also Han Confucianism, Neo-Confucianism, orthodoxy, and un-Confucian attitudes and ideas
Confucius, 5, 6, 11, 14, 19, 36, 40, 44n, 48, 72, 88n, 89, 91, 98, 99, 143, 153, 155n, 156, 160n, 161, 162, 167n, 177, 178, 196; ambivalence toward the relationship between Heaven and man, 19; *Analects*, 17, 48n, 89n; central ideas of, 14; discussion of inborn knowledge and learning, 163n; esoteric teachings of, 103n, 161n; ideal of legal prosecution, 115n; the worship of, 145
consultation, 115, 116
contentment, 197
contradiction, 20, 40; between adherence to the ideal and expedience, 43, 133, 139
conventions of the vulgar, 124
cooperation between the monarch and the ministers, 164
correlation, between breath and mental state, 159; between mind and words, words and deeds, likes-dislikes, praise-blame, and reward-punishment, 125; between spiritual essence and physical appearance, 154

INDEX-GLOSSARY

correspondence, between *ch'i* and *wu*, 154; between the cosmos and man, 36, 38, 88n, 153; between Heaven and Man, 13, 86, 127, 142–43; between Heaven and Man, uncertainties of, 87; natural, 177
corruption of the *Han-chi* text in the Sung dynasty, 68
corvée labor, 109
cosmic cycles, 155, 161n; numbers, 151, 157
cosmology, 10n, 20n, 21, 22, 26, 28, 35, 36, 40, 45, 53, 80, 86, 91, 103n, 112n, 151n, 153n; abnormal phenomena and disastrous occurences, 22n; anthropocentric, 12; celestrial cycle, 152; energy of spring, 159; dominant element and virtue of a ruling dynasty, 20n; numerological thinking, 112n; primitive notions of, 12n; systematization by Tsou Yen, 12n. *See also* correspondence between cosmos and man, etc.
court audience, 127, 143
creative thinking, 14n
criticism, likened to medicine, 173n; versus praise, 129
crossroads, 167
culture, 10n, 14
Custodian of Secret Archives, *see Mi-shu chien*
customary right versus governmental law, 137
Chao state, 84, 85
Chao Hsiang-tzu 趙襄子, 167n
Chang Ch'ien 張騫, 197
Chang Erh 張耳, 82
Chang Liang 張良, 15n, 81
Chang Ts'ang 張蒼, 20n, 134
chen 貞, 179n
Chen-hsing 鎮星, 151n
cheng 正, 112, 117

Cheng Mo 鄭默, 77
chi 幾, 184
chi 極, 135n
Chia I 賈誼, 15n, 21n, 133n
Chiang Hai 羌亥, 156n
chiao 郊, 141n
chiao 交, 195n
chiao-yao 僬僥, 155n
Chieh 桀, 167n
chien 簡, 135n
chien 檢, 197n
chien 鑑, 3, 38, 103, 106n, 163
Chien-an 建安, 33, 34, 43, 105n
chien-chia 檢柙, 164n
Chien-ch'a yü-shih 監察御史, 133n
Chien-yü-shih 監御史, 132, 133
chih 置, 169n
chih 至, 181n, 184, 185, 195n, 196n
chih 志, 30n, 44, 179n, 191, 192, 196n
chih 指, 旨, 7n
chih-chih 止之, 170n
chih-kung 致公, 186n
Chin, dynasty, 5, 180n; state, 116n, 152, 175n
Chin Mi-ti 金日磾, 172
chin-wen 今文, 28
ching 經, 11, 17, 103n, 183
ching 精, 159n
ching-ch'i 精氣, 86
ching-t'ien 井田, 92, 95, 139
chiu-p'in 九品, 192n
chiu-wen 舊文, 77
chou 州, 130n, 134n
Chou 紂, 167
Chou dynasty, 20n, 21n, 35, 87n, 93n, 103n, 104n, 116n, 120n, 128n, 132, 133n, 139n, 150n, 151n, 152n, 163, 168, 172n, 188n; breakdown of, 5, 8; concept of Heaven (*t'ien*) and its Mandate (*ming*), 19; old tradition of, 9
Chou-li 周禮, 99
Chou-mu 州牧, 126, 132, 133, 134n

INDEX-GLOSSARY

Chou-shu 周書, 63
Chu Po 朱博, 70
Chuang Tzu, 12, 40
chung 中, 112
Chung Chang 終張, 161n
Chung-ching 中經, 77
chung-jen 中人, 110n
chung-ku 中古, 196n
Chung-li I 鍾離意, 171
Chung-lü hsiao-wei 中壘校尉, 132n
chung-shih 終始, 161n
Chung-yung 中庸, 17
chün 郡, 130n, 134
chün-t'ien 均田, 139n
ch'an 讖, 28, 161n
Ch'en Chen 陳軫, 83n
Ch'en P'ing 陳平, 15n
Ch'en She 陳涉, 82, 83
Ch'en Yü 陳餘, 82
ch'en Yüeh 臣悅, 75
Ch'en-hsing 辰星, 151n
ch'en-t'u 塵塗, 179n
ch'eng 誠(成), 112, 189n
Ch'eng Ching 成荊, 156n
ch'eng-hsiang 丞相, 70, 71
ch'eng-men hsiao-wei 城門校尉, 132n
ch'i 氣, 154n, 189
Ch'i state, 169n
Ch'i-chih 七志, 77
ch'i-chü 起居, 149n
ch'i-chü chu 起居注, 149
ch'i-shu 氣數, 157n
ch'ien 乾, 191
ch'ien and *k'un* 乾, 坤, 144
ch'ih 耻, 49, 195–97
ch'in 親, 157
Ch'in dynasty, 15n, 16, 20, 21n, 22, 25, 29, 35, 70, 81, 82, 83, 84, 94, 132, 133n, 134n, 151, 163, 170, 171, 172n, 176; failure, 18; legalism, 14, 16; First Emperor, 13; empire, 13, 19; suppression of learning and proscription of books, 16, 145. *See also* legalism

ch'ing 卿, 130
ch'ing 頃, 94, 95
ch'ing 情, 44, 45, 47, 49, 64, 81, 82, 191, 192
Ch'ing dynasty, 53
Ch'ing Imperial Library Catalogue, 56
Ch'ing-chi 慶忌, 156n
ch'ing-hsin 情訊, 115n
ch'ing-i 清議, 3, 4, 31, 33, 34, 43; conflicting group interest, 32–33
ch'ing-t'an 清談, 50, 51, 180n
Ch'u state, 170, 176, 181
Ch'u state (of Hsiang Yü), 82–84
ch'üan 權, 86, 183
ch'üan 勸, 122n
ch'uang-nan 創難
ch'un 春, 159n
Ch'un-ch'iu, 16, 17, 103n, 148, 176

dates of the *Han-chi* and the *Shen-chien*, 73
death, 115
desire, 49, 172, 197. *See also yü* 欲
destiny, *see ming* 命
determination, 179
dialectic meanings, 104–105n, 125n
dialectical reversal, 122–23
dialectics, between dark and light, 125; between disorder and order, activity and quietude, etc., 128; change from benefit to harm, 168; change from submission to domination, 168; crime being not crime, 169; of the external and the internal, 195–96; of fortune and calamity, 173, 186; of yielding and unyielding, 171; "reaching does not reach", 182n; trouble mitigating trouble, 169
dialects of the Hsia and Ch'u peoples, 145
different conditions calling for different strategies, 82–84

212

different times calling for different strategies, 82–84
difficulties in employing the virtuous, 113
disasters and abnormal phenomena, see tsai-i
disasters and riots, 107, 122, 142
discipline, 197
discourse, see lun
discrepancy between knowing and realizing, 181n
"distant sacrifices" (wang 望), 142
divination, 150
Divine Intelligence, 109, 129, 180, 195, 196. See also shen-ming
"Doctrine of the Mean," see Chung-yung
dogmatism, 5, 8, 39, 50, 51, 52, 53, 90–91, 104–105, 143, 144, 145, 147–48, 152–53, 183, 184–85, 191
"Dragon-tail" 龍尾, 152
dualism, 12n
Duke Ching of Ch'i, 166n
Duke Huan of Ch'i, 111n, 165
Duke Hsi of Lu, 90
Duke of Chou, 188
Duke Shao of Chou, 168n
Duke Wen of Chin, 176
Duke Wen of Chu, 166
Duke Wu of Wei, 173
Duke Wu-ch'en of Shen, 176n

early Han court, 15n, 20n, 106n
eccentric nonconformity, 114
eclecticism, 14–15, 128–29, 136–37, 146, 152–53, 157–60, 162, 177–78. See also compromise
eclipse of the sun, 86, 143
eclipse of the sun and the moon, 142
education, 20, 21, 41, 50, 89, 91, 96–99, 108n, 136, 148, 188n, 192–93; and culture, 14; and punishment, 135–36; Confucian, 99, 110; efficacy, 89; the metaphor of metallurgical brazier and water-pump, 120; vain instruction, 136
egalitarian system, 95
emotions, 169, 172, 187–89. See also feelings, ch'ing 情
emperor, see ruler
Emperor Ai, 95, 171n
Emperor Chao, 174n
Emperor Ching, 173
Emperor Ch'eng, 171n
Emperor Hsien, 3, 33, 34, 77, 78
Emperor Hsüan, 96, 172n
Emperor Kao-tsu, 15n, 95, 96, 170, 171n, 172, 173. See also Liu Pang
Emperor Kuang-wu, 26, 27n, 95, 171, 172. See also Liu Hsiu
Emperor Ming, 171
Emperor P'ing, 27n
Emperor T'ai-tsung of T'ang dynasty, 55
Emperor Wen, 15n, 16, 21n, 23n, 92, 94, 96, 109n, 115n, 134n
Emperor Wu, 15n, 17, 21n, 23, 95, 96, 131, 161n, 172n, 173, 174n
Emperor Yuan, 96, 171n
empire, 14, 122; as a public institution, 41; the reality of, 52. See also state
Empress Ming-te, 174
Empress née Chao, 171n
Empress née Lü, 86, 173n
Empress née Ma, 149
Empress Ting-t'ao, née Fu, 171
environments, 155
equality, 93
erh-i 二儀, 144n
erh-shu 二豎, 165n
eunuch, 25n, 31, 96
evil, 182; deeds, 194; nature, 193
exemplar of Hsün Yüeh's original Han-chi lun, 71
exemplary women of the past, 144
expediency, 86, 116, 133, 146–47, 183. See also ch'üan 權

INDEX-GLOSSARY

expertise in government affairs, 15
external (versus the internal), 18, 44, 46, 51, 175; attraction (temptation), 190; evidence, 115n

fa 法, 11
falsehood, 107, 109, 121, 129
fame, 183
family morality, 137n
Fan Li 范蠡, 168n
Fan Yeh 范曄, 74
fate, 27, 29, 88, 155. *See also ming* 命
favorite concubine of a ruler, 165n, 173
favoritism, 165
feelings, 43, 46, 47, 48n, 110, 115, 189–91; dichotomy between human nature and, 49; discrimination between the finer and the cruder forms, 50; emphasis on, 45–46; higher and lower forms, 49; important role in character formation and personality integration, 49; issues on, 49; likes and dislikes, 41, 47, 50. *See also ch'ing* 情
feminine rites and virtues, 144
feng-shan 封禪, 141n
feudalism: self-styled, 95; feudal lords, 82n, 94, 133, 138, 148; feudal society, 139n; fiefs, 94, 138, 139n; self-designated, 139
firmness, 180, 186
Five Classics, 16n, 17, 103, 161n. *See also ching* 經
Five Elements, 12n, 20n, 21n, 87n, 88, 97, 143n, 151n. *See also* cosmology
Five Lessons, 108
Five Planets, 151
five programs of government, 107, 108, 112
five virtues, 106n, 107
flexibility, 86, 95, 97–98, 121, 127, 134–38, 140–42, 170–72, 174. *See also* expediency
flexibility and tolerance, 39–40, 43
Four Criminals, 188
"four directions," 12n
Four Elders of Mount Shang, 171, 197
four evils, 112
fu 復, 194n
Fu Yüeh 傅說, 173n
fu-jung 婦容, 144n
fu-kung 婦功, 144n
fu-te 婦德, 144n
fu-yen 婦言, 144n

geomancy, 143n
Golden Age, 95n
golden mean, 112, 157, 158, 177. *See also chung*
good and bad, 29n, 109, 111, 189, 192, 193
good and bad fortune, 186–87
good conduct, 181
good counsel, 170
good faith, 195
governing, the act of theft and robbery, 122; a burden to the Sage-ruler, a pleasure to the mediocre ruler, 122; chicken-herding as a metaphor, 121; fishing as a metaphor, reining the horse as a metaphor, 121; metal as a metaphor, water as a metaphor, 119; river crossing as metaphor, 120
government, pivots of, 111; primary aim of, 96
gradualism, 98, 135–36, 142
Grand Mandate, *see ta-ming*

Han army, 84–86
Han Confucianism, 10n, 20n, 22n, 31, 40, 52; advocating change of dynasty, 24; ambivalence toward human reason, 19; Ancient Texts

School, 28; contradiction to reality of dynastic rule, 20; different reaction towards Wang Mang's failure, ambivalence toward the ruling regime, 26; the impact of Wang Mang's failure, 25; in disarray, 26; diversity of the orthodoxy, 17–18; dogmas of, 36; emphasis on historical learning, 16–17; historical movement, 18; Modern Texts School, 28, 145 orthodox, 14, 51; protest, 31; reforms, 16–17; relationship with imperial ideology, 17; success of moderate reform, 23; syncretism, 14; triumph, 14. *See also* Confucianism

Han dynasty, denunciation of the Ch'in, 21n; effort to recover ancient texts, 145

Han Hsin 韓信, 84, 85

Han Fei, 7, 11, 12, 13

Han imperial family, 27n

Han rulers, achievements of, 72, 96

Han state, 83n

Han tombs, 15n

Han-chi hsü 漢紀序, 75

Han-chi mu-lu 漢紀目錄, 76

Hao-jan 浩然, 195n

harmony, 157, 158, 174; between body and spirit, 160; contrast to conformity, 178; in diet, 177; of action, 177–78; of counsels, 177; of sounds, 177; of spirit, 179. *See also ho*

Heaven, 19, 22, 36, 48; God, 5; complementary interaction with Earth, 164; Mandate of, 20, 22, 23, 24, 25, 27. *See also t'ien-ming*; spiritual essence of, 106; the will of, 6, 19. *See also t'ien*

Heaven, Earth and Man, 106

historical circumstances, 36

historical events, 50

historical learning, 53

historical records, 40, 42, 43, 148

historical reflections, 38. *See chien* 鑑

history, 11, 12, 26, 36, 38, 74, 80, 163; examples from, 51, 197; the function of 148; lessons of, 14, 36, 50, 81, 103n, 104; importance of written records, 75; the onslaught of, 13; relationship with cosmology, 22; Taoist view of, 15. *See also chien* 鑑

history and culture, 18

ho 和, 112, 113

Ho Yung 何顒, 34

honesty, 126–27, 129

Hou-chi 后稷, 141n

Hsia dynasty, 93n, 105n, 128n, 163, 168, 176, 188n

Hsia Yü 夏育, 156

Hsiang Yü 項羽, 22n, 81, 82, 83, 84, 86

Hsiao Hsün-tzu 小荀子, 59, 60, 61, 63

hsiao-tz'u 爻辭, 191

Hsiao-wei 校尉, 131–32

hsien 仙, 155

hsien 賢, 31n

hsien 縣, 133n, 142n

hsien-jen 先人, 194n

hsien-kuan 縣官, 142n

hsien-ti 先帝, 76

hsien-tien 限田, 139n

Hsien-yang, 151

hsin 心, 44, 191, 192

hsing 性, 29n, 44, 45, 47, 48, 49, 64, 92, 128, 156, 181n, 186, 191–92

hsing 行, 44, 181n

hsing 形, 81, 189n

hsing-yung 形容, 154n

hsiu 休, 186n

Hsiung-nu, 172n

Hsü 許, 33

hsü-ch'ü 虛取, 122n

hsü-yü 虛與, 122n

hsüan 玄, 185

INDEX-GLOSSARY

hsüan-hsüeh 玄學, 4
hsüan-tu 懸度, 167
Hsün clan, 34, 52
Hsün Shuang 荀爽, 33, 34, 103n, 161, 193n
Hsün Tzu, 7, 10, 11, 12, 13, 16n, 17, 19, 20, 36, 47n, 140n, 181n, 187
Hsün Yü 荀彧, 34
Hu-ch'en 虎臣, 104n
Hu-pen hsiao-wei 虎賁校尉, 132n
hua 化, 10n, 30n
Huan T'an 桓譚, 28
Huang Chen 黃震, 48, 59
huang-pai 黃白, 160n
Hui Shih 惠施, 8n
human efficacy, 11, 12, 13, 18, 20, 35, 88–89; action and reason, 9; in doubt, 26, 37; pragmatic, 6–8
human efforts, 9n, 29n–30n; 37, 45, 48, 128n, 155, 187
human existence in the universe compared to a traveller's stay at an inn, 152n
human feelings, 4, 48. *See also* feelings
human morality, 50, 51
human nature, 10, 30n, 44, 47n, 80, 128, 181n, 186, 187–92, 194. *See also hsing* 性
human nature and feelings, meat and wind as metaphor of, 190
human predicament, 35
human sacrifice, 166
human understanding, 39
humanity, "inborn" moral and intellectual qualities, 29; moral intention, determination and effort, 29. *See also jen* 仁
hundred schools of philosophers, 129, 185
Hung-fan 洪範, 87
hypocricy, 121

i 義, 47, 103, 106, 137, 189, 190, 195, 197

i 意, 191, 192, 194n
I Ti 儀狄, 176n
I-ching, 17, 71, 87, 90, 92, 103n, 119n, 127n, 143, 147, 150n, 187, 191, 192, 193n
i-i 泄泄, 163n
ideal and reality, 51, 93
Imperial Academy, 17, 145. *See also T'ai-hsüeh*
imperium, 12, 18, 35, 42, 143n
inanimated and animated stages of human mind, 48
inconsistency, 38
inferences, 161
influence of different enviornments, 128
influence of different governments, 128
influence of different times, 128
innate nature, *see hsing* 性
inner and outer dimensions of human existence and morality, 44
inner and outer strength, 51
inner conflict between the good and bad, 49, 192
"inner defense," 44, 177
inner harmony versus outer harmony, 153n
inner motives, 44, 115n
inner nature, nourishing of, 157
inner palace, 117n, 127n, 149
inner virtue of man, 30, 150; in contrast to outside influence, 152n
inner visions, 158
inner vis-á-vis the outer realms, 35, 44, 47, 49, 50, 53, 152
inner world, 29, 35n, 44, 48
instinctual desires, 49
intention and determination, 44, 180. *See also hsin* and *chih* 志
internal and external factors, 45, 114
internal bandits, 177
internal treasure, 177
inward and outward search, 51, 182

216

INDEX-GLOSSARY

Jan Keng 冉耕, 88, 90
jen 仁, 20, 48, 103, 104, 106, 166, 189, 190
jen 任, 165n
jen-chih 人豕, 173n
jen-chün 人君, 31n
jen-i 仁義, 46
jen-lun 仁論, 196n
Ju-chia 儒家, 53n. See also Confucianism
justice, 93, 112. See also *cheng, i* 義

kai 概, 197n
Kai K'uang-jao 蓋寬饒, 24, 25n
kao-huang 膏肓, 89, 165
kao-t'ien 告天, 141n
Kao-yao 皐陶, 109
kang 剛, 106n
kindheartedness, 117. See also *jen* 仁
King Chou of Shang dynasty, 188
King Hsüan of Chou dynasty, 87, 90, 168n
King Hsüan of Ch'i, 175
King Hui of Ch'in, 83n
King Kao-tsung of Shang dynasty, 173
King Kou-chien of Yüeh, 168, 173
King Kuai of Yen, 168n
King Kung of Ch'u, 175–76
King of Han (Liu Pang), 81–85
King of Wu, 173n
King Shao-k'ang of Hsia dynasty, 168
King Shun, 105n, 106n, 109n, 112n, 133n, 147n, 166–67, 171n, 177n, 182, 188, 186n
King T'ang of Shang dynasty, 166n
King Wen of Chou dynasty, 188
King Wu of Chou dynasty, 104n
King Wu of Ch'in, 156n
King Wu-ting of Shang dynasty, 168n, 173n
King Yao, 87n, 90, 106n, 112n, 147n, 166–67, 171n, 176, 188,

196n
King Yü of Hsia dynasty, 105n, 156n, 158, 176, 182
knowledge, 9n, 40, 126, 129, 163, 184; approximation of truth, 37; cosmological, 37; historical, 36–37; knowing oneself versus knowing others, 182, 183n; limitation of, 37; reasoning from the concrete to the abstract, 37–40, 90; truth, 4, 35–40, 91
ko 格, 104n
Ku Yung 谷永, 24, 25n
ku-wen 古文, 28
kua 寡, 163n
Kuan Chung 管仲, 111n, 165
Kuei Hung 眭弘, 24
Kuei Yu-kuang 歸有光, 60n
kuei-mang 桂莽, 155n
kung 公, 31n, 112, 130
kung and *ch'ing* 公卿, 126
kung-chu 公主, 149n
Kung-sun 公孫, 187
Kung-sun Ch'en 公孫臣, 21n
Kung-sun Ch'iao 公孫僑, 116n
kung-t'ien 公田, 93n
kung-wang 公王, 149n
Kung-yang tradition, 90n
k'ao-shih 考試, 126n

land ownership, 93, 127, 138–39; restriction of, 92, 95
land policy, 92–96, 138–39
land tax, 92n, 93n
landlordism, 92n, 94, 95
language problem, 7n, 185
Lao-tzu, 7n, 127n, 186n
law, 11, 18, 21n, 41, 42, 96, 97, 99, 108, 111, 114, 135n, 136, 137, 148, 188n, 192, 193. See also *fa*
law and education, 96–99, 106–107
law of change, 99
law of nature, 88, 91, 154, 187n. See also the Way

learning, 99n, 146, 163
legal investigation, 115n
legal restraint, 50, 97
legalism, 11, 12, 13–14, 15, 16n, 17, 19, 188n, 194n, 195n. *See also* Ch'in dynasty
leniency, 116
li 禮, 9, 11, 99, 103n, 186n, 194n
li 理, 88, 121n, 153n, 187
li 利, 47
li 力, 157
Li Hsin 李尋, 72
Li I-chi 酈食其, 81n
Li T'ao 李燾, 56
li-chiao 禮教, 4, 108n
li-shih 吏事, 15, 28
life, 29n, 115, 118, 152n, 155, 159, 160, 181n
life-nourishing, 108–109, 157–60
likes and dislikes, 108, 109, 118, 124, 144, 189–90. *See also* feelings
litigations, 115n
Liu Hsiang 劉向, 24, 25, 45, 77, 187, 188
Liu Hsiu 劉秀, 26. *See also* Emperor Kuang-wu
Liu Ling 劉伶, 152n
Liu Pang 劉邦, 81–85. *See also* Emperor Kao-tsu
Liu-li 六沴, 143n
Liu-tzu 六子, 60
Lo-yang, 33
low tax rate, 92n
Lu Wen-shu 路溫舒, 15n
luan-cheng 亂政, 104n
luan ch'en 亂臣, 104n
lun 論, 36, 38, 55, 69, 80, 81, 92, 96
lüeh 署, 135n

magic, absolution from illness and danger, 154; art of immortality, 155; channeling the body's energies, 158n; healing illness, 158–59; of immortals, 157; for longevity, 156; nurturing the spirit, 158, 160; preserving vital energies, 157
magistrate of the district, 130
Mandate, *see ming* 命
market transactions, 141. *See also* commerce
Marquis of Chin, 165n
masses, the, 114, 115, 121, 140, 183. *See also* the people
Master Chu of T'ao 陶朱, 168
Master Chuang of Pien 卞莊子, 83
meaning and things, 6n
medicine, 159–60, 165, 173
meditation, 159n
Mencius, 6, 10n, 16n, 17, 19, 20n, 47n, 48, 93n, 95n, 109n, 146n, 166, 180n, 181n, 187
Meng Pen 孟賁, 156
merchants, 92n
meritocracy, 20, 23, 32, 126, 129–30
metamorphosis, 157
metaphor: "meat and wine", 46–49
mi 密, 135n
Mi-shu chien 祕書監, 3, 34, 75, 77, 78
military, 108, 131, 132
military art, 126
military merits, 97, 131
military preparedness, 111
militia training, 131–32
mind, 194. *See also hsin*
ming 明, 179n
ming 命, 19, 29, 30n, 37, 44, 45, 47, 150n, 155n, 156, 160, 182n, 185, 186, 187, 191
Ming dynasty, 53, 54, 134n
Ming-chia 名家, 6, 7n, 39
ming-chiao 名教, 4
ming-chün 明君, 31n
Minister of Crime, 109
ministers, 107, 114, 118, 123, 124, 170, 174; dilemma of, 169; obligation of, 169; six types of, 72
mirror, 38, 103n, 163. *See also chien* 鑑
Mo Tzu (Ti), 6, 153

INDEX-GLOSSARY

monarch, *see* ruler
monetary system, 127, 139–41
Monitor-in-Attendance, 164
moral action, 44. See also *hsing* 行
moral autonomy, 47–48
moral courage, 31
moral decision, 190
moral education, 41, 44, 116
moral influence, 126
moral integrity, 124
moral judgment, 40, 188n; on accomplished actions, 46
moral situation, 44
moral standards, 11n, 18, 127
morality, 32, 35, 40, 177
motives, 194
mou 畝, 94
mu 牧, 140
music, 103, 116, 118, 174, 177, 178n
Mystical Learning, *see hsüan-hsüeh*
mysticism, 27–28

name, 39, 192; language and ideas, 10n; Taoist distrust of, 7n. *See also ming* 名, *ming-chia*, nominalism
name and conscience, 153
name and reality, 6n, 109, 192
name and substance, 162
Nan-chih-wei 南之威, 176
natural versus unnatural, 161
nature, 19, 29n, 30n, 108, 112n, 119n, 150n, 153n, 156, 185, 186, 191; Taoist view of, 188n; and fate, 89, 90; and man, 86
nei-shih 內史, 144, 149
neo-Confucian critique of Hsün Yüeh, 48–49, 63; identification of *li* (reason) and *hsing* (human nature), 187n
neo-Confucianism, 19n, 47, 49, 50, 53, 153n
nine types of people, 192
nominalism, 6, 39. *See also* name
non-action, 14–15, 112, 121

number-mysticism, 119, 150n. *See also* cosmology

observable and non-observable, the, 45–49, 189. *See also* inner and outer
official emoluments, 127, 138
Official Erudities, 17, 127, 131, 145. *See also po-shih*
omens, 27, 86, 87, 90, 142, 149, 150, 152, 153, 160, 161n, 168, 177
optimism, 11, 18, 35, 87, 103–105, 107, 109–110, 112
ordering and harmonizing the world, 29
Ordinance for Feminine Education, 144
ordinary human being, 27n, 41, 44, 46, 50, 51, 110, 120, 174. *See also chung-jen*, the people
orthodox teachings inside the palace, 127, 144
orthodoxy, 14, 28, 52, 53, 103n, 108n, 129. *See also* Confucianism
outer court, 117n, 127n
outer decorum, 44
outer defense, 44
outer sphere, 29, 150n
outside influence, 44, 47, 128–29, 152–53, 176–77, 190–91, 192–93
over-population, 95

pa-chi 八及, 32
pa-ch'u 八廚, 32
pa-ku 八顧, 32
pa-tsün 八俊, 32
Pan *Chieh-yü* 班婕妤, 174
Pan Ku 班固, 192n
Pan Piao 班彪, 27
pao 報, 123n
Pao-chang 保章, 143n
pao-pien 襃貶, 3
pardon, five occasions for, 116. *See also* amnesty

219

INDEX-GLOSSARY

"pass," 關, 159
peasants, 92n. *See also* masses, people
Pei-chün chung-hou 北軍中候, 132n
people, 31n, 41, 42, 43, 107, 108–109, 110, 118, 119, 121, 122, 123, 128, 133, 136, 141, 166, 193; the realm belongs to, 25; water as a metaphor
personal conscience, 43
preversion of official prerogatives, 124
pessimism, 35n, 46, 51, 186n, 195n, 197n
petty officials, 113–14
petty man, 110–11
philosophic works, *see tzu*
physiognomy, 154
pieh-chi 別集, 59
Pieh-lu 別錄, 77
pien 變, 155n
Pien Ch'üeh 扁鵲, 89
pien-ch'an 諞譔, 161n
pien-shu 變數, 112n
Ping Chi 丙吉, 172
Po-i 伯夷, 120n, 196n
po-shih 博士, 17, 145
population, 127, 134
positivism, 8
possible and the impossible, the, 90
"praise and blame," 3, 50, 94n, 109, *see pao-pien*
praying, 153–54; for rain, 166n
pre-Sung quotations of *Shen-chien*, 65–67
pre-T'ang periods, 73
Prince of Kuang-ch'uan, 175
Prince of Ting-t'ao, 171n
private levies, 119
profitseeking, 41, 129, 190; mistaken for public-spiritedness, 114
provincial government (*chou*), 132–33
pu-ai 不愛, 165n
Pu-ping hsiao-wei 步兵校尉, 132n

public, the, 41–42; (official) versus private (personal), 119, 138
public-spiritedness, 165, 174, 175
punishment, 110, 115, 116, 130, 140; by mutilation, 115, 126, 134–35; cruel, 136. *See also* reward and punishment
pure conversation, *see ch'ing-t'an*
pure criticism, *see ch'ing-i*
P'ang-keng 盤庚, 127n
P'eng-ch'eng 彭城, 84–85
P'eng-tsu 彭祖, 156n

realism, 98, 136
reality, 125; complicated and dialectical structure of, 39; and ideal, 51; and name, 109; and truth, 39
reciprocity, 118, 119n, 177; between the ruling and the ruled, 122–23
reconstruction of Hsün Yüeh's *lun* (discourses) in the *Han-chi*, 69–72
rectification of names, 39, 192
rectifying the self, 107, 117, 121; the social conventions, 110
reflection, 3, 103n, 163. *See chien* 鑑
reform, 23n, 28, 29, 143n
regional power, 133–34n
relationship between nature and man, 37; between *ts'ai* and *hsing*, 180n
remonstration, 114, 170, 182
responses to stimuli, 191
restoration, of Chou dynasty, 168n; of Yen state, 168–69n; of Yüeh state, 168n
reversal of political fortune, 123
reward and punishment, 108, 110, 111, 129
reward for military merit, 131
righteousness, 41, 47, 103, 117, 165, 170, 171n. *See also i* 義
rites and rituals, 21n, 108, 110, 113, 114, 118, 124, 132, 142, 143, 154, 164, 174; of ploughing the royal

220

INDEX-GLOSSARY

field, 109; of raising silkworms, 109
ruler, 23, 50, 51, 53, 105, 107, 108, 112, 114, 116–17, 118, 119, 124, 162, 169, 174–75; daily routine, 116; dilemma of, 169; employing the virtuous, 107; favorite women of, 115; Guardians and Tutors of, 116n; personal feelings of, 41–43; relatives of, 31; selfless, 41; serving the people; six types of, 72; versus the Sage, 50
ruler and the ministers, 42, 113, 114, 169, 175
rulership, 40–44, 123, 164–67, 170–73. *See also* ruler

sacrificial offerings, 12n, 141n, 142, 154
Sage, becoming emperor, 25; versus the ruler, 50
Sagehood, 48
San-chün 三君, 32
San-kung 三公, 70, 71
School of Names, *see ming-chia*
self-assertive, 183
self-improvement, 29, 90n, 129
self-indulgence, 175
self-interest, 175
self-knowledge, 51, 182–83
self-mastery, 47, 49, 51, 172, 173, 180
self-relection, 197
self-sacrifice, 42, 166
selfishness, 46, 107, 130
shame, 49, 195–97. *See also ch'ih*
Shang dynasty, 87n, 93n, 103n, 104n, 109n, 127n, 128n, 150n, 151n, 153n, 163, 166n, 167n, 168n, 173n, 188n; oracle scripts, 12n
Shang-chu 尚主, 127, 147–48
Shang-hsia hsiang-yü 上下相與, 122n
shang-ku 上古, 196n
Shang-shu 尚書, 148, 149

Shantung, 84
Shao-fu 少府, 117n
Shao-k'ang 少康, 168n
She-sheng hsiao-wei 射聲校尉, 132n
Shen 申, 3, 38, 104n
shen 神, 157, 179, 184
shen-ch'i 神氣, 90, 154n
shen-hsien 神仙, 155n
shen-hsin 神心, 27
shen-ming 神明, 179n, 195n
sheng 生, 156n. *See also* life
Sheng 省, 134n
sheng-chiang 升降, 193n
Shih 詩, 103n
shih 時, 93
Shih 史, 148, 149
shih 實, 153n
shih 市, 122n
shih 士, 109n
shih 事, 148n
shih 勢, 81, 82, 83, 84, 88, 89, 92, 97
Shih Hsien 石顯, 96
Shih-chung 侍中, 74, 75
Shih-erh-tzu 十二子, 60
shih-jen 市人, 122n
shou kung ling 守官令, 34
Shu 蜀, 33
shu 數, 88, 119n, 151n, 155n, 182n, 185n
shu 恕, 117
Shu Hai 豎亥, 156n
Shu-ching, 17, 87n, 103n, 104n, 106n, 116n, 173n
similar stategy producing different results, 84–86
similar tasks in different *shih*, 84
similar tasks of different *ch'ing*, 86
simple-mindedness, 39, 90
sincerity, 109, 112, 129, 180, *see ch'eng*
Six Dynasties, 55, 181n
six tempers, 107
skepticism, 4–9, 12, 37, 50, 51, 53, 86–88, 182–83, 186, 187–88, 194–95

221

society, 40; corrupt, 31; the supremacy of, 11; traditions of, 18, 129, 183, 195; and culture, 9, 11, 19
sophists, 6, 8, 39, see ming-chia
sovereign, see ruler
speech, 184, 185, 192
spirit, 38, 141, 153, 189
spiritual mind, see shen-hsin
Spring-and-Autumn, 90n, 94, 99, 181n, see Ch'un-ch'iu
Ssu-k'ung 司空, 34
Ssu-ma 司馬, 131n
Ssu-ma Ch'ien 司馬遷, 22n
Ssu-ma Kuang 司馬光, 54, 72, 74
Ssu-ma's Military Arts, 131
Ssu-tu 四瀆, 142n
Ssu-t'u 司徒, 109n
standard test, measuring the open field as a metaphor, sailing and sinking of boat as a metaphor, 129–30
state, 11, 13, 41, 42; affairs of, 116; approaching extinction, 114; archives of, 148–49; four evils of, 107, 108; holistic, 118, 140; imperial, 18; in danger, 114; in decline, 113; in good order, 113; nine conditions of, 113–15; the power and reason of, 11
state and society, 35
state cult, simplification of, 140–41
statecraft, see li-shih
strategy, 81, 82, 86, 162
strengthening the trunk (central government) and weakening the branches (regional power), 133
stricture on petty men, 72
Su Ching 蘇竟, 27n
Su Wu 蘇武, 172, 197
su-chih 素志, 167n
su-ssu 素絲, 167n
su-wang 素王, 167n
submission, proper and improper, 124

subversive teaching, 54
Sui dynasty, 134n
Sui-hsing 歲星, 151n
Sung dynasty, 52, 53, 54, 56, 57, 58, 73n, 78n
Sung I 宋義, 83
Sung official catalogues, 55, 59
superfluous offices, 138
superior man, 37, 80, 109, 110, 136, 157, 163, 164, 167, 174, 177–79, 181, 183, 186, 194–96
superiority of spirit over matter, 179
superlative, the, 181n. See also chih 至
superlative wisdom, see tu-chih
superstitions, 40, 129, 150–51
sympathy for the criminal, 115
system of selection and promotion of officials under the Wei dynasty, 193n

ta 達, 179n, 185n
ta-ming 大命, 20
Ta-ssu-k'ung 大司空, 69
Ta-ssu-nung 大司農, 117n
ta-t'ung 大同, 20
tang-ku 黨錮, 32, 43
tao 道, 9, 134n. See also the Way; likened to a road or a river, 27n
Taoism, 4, 16n, 17, 159, 189n; art of longevity, 158n, 160; breathing and sexual techniques, 158n, 159; distrust of names, 7n; Huang-Lao School of, 8, 15n; non-action and quietism, 14, 15, 135n; positivist aspect of, 9n; sentimentalism, 49n; view of nature, 188n. See also magic
Taoists, 6, 8, 12, 21, 52, 152n, 179n, 188n
Taoist adviser at the early Han court, 81
taxation, 92n, 119. See also land ownership

INDEX-GLOSSARY

te 德, 51, 157
temperance, 160
tension between dogmatism and skepticism, 35, 43
textual relation between *Han-chi* and *Shen-chien*, 64–65, 70, 72–73
Three Ancient Dynasties: Hsia, Shang, Chou, 16, 94, 105
three grades of men, 44, 155. *See also chung-jen*
Three luminaries: sun, moon, and star, 151
three primal spheres: Heaven, Earth, and Man, 88, 90, 107
Ti 狄, 167n
tien 典, 105n. *See also ching* 經
Tiger Minister, 104n, 105n. *See also hu-ch'en*
time of emergency, 144
timeliness, 118
training, 156
Treatise of Sacrifices, 141n
Treatise on the Five Elements, 22n, 86
truth, 11n, 39; approximation of, 51; relationship between idea and reality, 39. *See also* knowledge
tu-chih 獨智, 27
Tung Chung-shu 董仲舒, 22n, 23n, 24, 45, 49, 95
Tung Hsüan 董宣, 171n
t'ai-ho 太和, 20
T'ai-hsüeh 太學, 17
T'ai-po 太白, 151n
t'ai-p'ing 太平, 20, 23, 25
T'ai-wei 太尉, 70, 132
T'ang (legendary ancient dynasty), 90. *See also* King Yao
T'ang dynasty, 53, 55, 73, 133n, 134, 139n
T'i 體, 184n
t'i-tz'u 題辭, 59, 62
t'ien 天, 19, 22, 24, 25, 27. *See also* Heaven

t'ien-ming 天命, 25. *See also* Heaven
t'u-ch'an 圖讖, 27n
t'uan 彖, 191
t'ui 推, 197n
T'un-ch'i hsiao-wei 屯騎校尉, 132n
t'ung 通, 112
t'ung 同, 195n
t'ung-kuan 彤管, 144n
tsai-i 災異, 22, 27
tse 責, 194n
tse-pei 責備, 136n
tsi 積, 9n
Tsou Yen 鄒衍, 7, 12, 13
tsu 祖 and *tsung* 宗, 76
ts'ai 才, 材, 45, 180, 181n, 182
Ts'ao P'i 曹丕, 33
Ts'ao Ts'an 曹參, 15n
Ts'ao Ts'ao 曹操, 33, 34, 105n, 141n, 174
Ts'ao-Wei regime, 77
tzu 子, 3, 16n, 53n, 60
Tzu Hsia (Pu Shang) 子夏 (卜商), 89n
Tzu-chih t'ung-chien, 54, 72, 74
tzu-jan 自然, 4

ultimate, the, 13
Ultimate Harmony, *see t'ai-ho*
uncertainty in the transmission of the *Shen-chien*, 59
uncertainty of the transmission of Hsün Yüeh's writings during the Sung, 68
un-Confucian attitude and ideas, 5, 46, 51, 52
unification of China, 11
union with the Way, 195
union with all under Heaven, 195
unity: human, 6; of the moral, sociopolitical, and cosmic orders, 20, 25, 30; of the "inner" and the "outer" worlds, 30, 31
unity and diversity, 178
universal military service, 131n

INDEX-GLOSSARY

Universal Peace and Equality, see *t'ai-p'ing*
universe, likened to man's abode, 152
unknown, the, 8, 13, 27
unobscrvable inner world of men, 50. See also the observable and non-observable
unorthodox cults, 129

"vain trade," 122n
vengeance, 136
versatility, 88; the metaphor of the net and the bird, 146
virtue, 179, 194, 197. See also moral
Viscount of Chao, 168
vital energies, water as a metaphor of, 158
vulgar customs, 124n, 195

waiting strategy, 83–84
wang 望, 142n
Wang Chien 王儉, 77
Wang Chih 王銍, 56
Wang Ch'ung 王充, 29, 30, 31, 35, 36, 37, 44, 45, 48, 50, 86
Wang Fu 王符, 30, 31n, 35, 44, 48, 50
Wang Mang 王莽, 25, 26, 29, 171
warlords, 132n
Warring States, 8, 9, 16, 83, 84, 181n
water element and virtue, 20n
Way, the, 42, 43; anthropocentric view of, 106n; the essence of, 184; foundation of, 103; fruits of, 112n; of Heaven, 8, 37, 38; of King Yao, 167; of man, 8, 37; of moral integrity, 194; of superior man, 144; of wicked rulers, 167n; substance of, 112; "warp" of, 103n, 112n. See also *tao*
wei 緯, 28, 40, 103n, 161–62
wei 唯, 謂, 7n
Wei dynasty, 4, 33, 105n, 152n, 180n

Wei state, 83n
well-field system, the, 92, 95, 139, see *ching-t'ien*
wen 文, 153n
wen-li 文吏, 30n
wen-wen 汶汶, 163n
Western Regions, 197n
white horse, 6n
will power, 45, 47, 49, 50
written records, 146. See also historical records, history
wu 物, 7n, 154n, 189n
Wu state, 33
wu-ch'ing 無情, 115n
Wu-hsing hsiang-k'e 五行相克, 21n
Wu-hsing hsiang-sheng 五行相生, 21n
wu-kung 武功, 131n
wu-shih 五事, 107n
wu-wu 戊午, 69
wu-ying 五營, 132n
wu-yüeh 五嶽, 142n

Yang Chu 楊朱, 167
Yang Hsiung 揚雄, 10n, 27, 64, 187
Yellow Turbans uprising, 32, 34, 132n
Yen Hui 顏回, 88, 90
Yen-tzu 晏子, 166n, 178
yin 引, 170n
yin and *yang*, 12n, 86, 90, 106, 112n, 144, 152, 159, 189n, 193
Yin dynasty, 127n, 151, 168. See also Shang dynasty
Yin-chien 殷鑑, 103n
yin-li 陰禮, 144n
yin-yang and Five Elements Cosmology, the origin of, 12; mechanistic system, 13
yin-yang chih-li 陰陽之禮, 144n
Yin-yang School, 161n
yin-yao 陰葯, 160n
ying 應, 7n
Ying-ch'uan 潁川, 33, 34, 74
ying-hsü 盈虛, 127n

ying-huo 熒惑, 151n
yu 柔, 106n
Yu ssu-ma 右司馬, 176n
yu-yu 悠悠, 163n
Yuan Hung 袁宏, 54, 74
Yuan Shao 袁紹, 34
yüan-ch'en 元辰, 152n
yü 遇, 90, 97
yü 欲, 49

yü 與, 122n
Yü Chung-yung 余仲容, 65n
Yü-shih ta-fu 御史大夫, 69, 70
yüeh 約, 146n, 184n
Yüeh Kuang 樂廣, 182n
Yüeh state, 9n, 168
Yüeh-ch'i hsiao-wei 越騎校尉, 132n
yün 運, 155n
yün-shu 運數, 155n

Library of Congress Cataloging in Publication Data

Hsün, Yüeh, 148-209.
    Hsün Yüeh and the mind of Late Han China.

    (Princeton library of Asian translations)
    Bibliography: p.
    Includes index.
    1. Ethics, Chinese.   I. Title.   II. Series.
BJ117.H8413   181'.09512   79-3196
ISBN 0-691-05292-6